Studies in Systems, Decision and Control

Volume 268

The series "Studies in Systems, Decision and Control" (SSDC) covers both new developments and advances, as well as the state of the art, in the various areas of broadly perceived systems, decision making and control–quickly, up to date and with a high quality. The intent is to cover the theory, applications, and perspectives on the state of the art and future developments relevant to systems, decision making, control, complex processes and related areas, as embedded in the fields of engineering, computer science, physics, economics, social and life sciences, as well as the paradigms and methodologies behind them. The series contains monographs, textbooks, lecture notes and edited volumes in systems, decision making and control spanning the areas of Cyber-Physical Systems, Autonomous Systems, Sensor Networks, Control Systems, Energy Systems, Automotive Systems, Biological Systems, Vehicular Networking and Connected Vehicles, Aerospace Systems, Automation, Manufacturing, Smart Grids, Nonlinear Systems, Power Systems, Robotics, Social Systems, Economic Systems and other. Of particular value to both the contributors and the readership are the short publication timeframe and the world-wide distribution and exposure which enable both a wide and rapid dissemination of research output.

** Indexing: The books of this series are submitted to ISI, SCOPUS, DBLP, Ulrichs, MathSciNet, Current Mathematical Publications, Mathematical Reviews, Zentralblatt Math: MetaPress and Springerlink.

More information about this series at http://www.springer.com/series/13304

Shanling Dong · Zheng-Guang Wu · Peng Shi

Control and Filtering of Fuzzy Systems with Switched Parameters

Shanling Dong
National Laboratory of Industrial
Control Technology
Institute of Cyber-Systems
and Control, Zhejiang University
Hangzhou, Zhejiang, China

Zheng-Guang Wu
National Laboratory of Industrial
Control Technology
Institute of Cyber-Systems
and Control, Zhejiang University
Hangzhou, Zhejiang, China

Peng Shi
School of Electrical
and Electronic Engineering
University of Adelaide
Adelaide, SA, Australia

Victoria University
Melbourne, VIC, Australia

ISSN 2198-4182 ISSN 2198-4190 (electronic)
Studies in Systems, Decision and Control
ISBN 978-3-030-35568-5 ISBN 978-3-030-35566-1 (eBook)
https://doi.org/10.1007/978-3-030-35566-1

This Springer imprint is published by the registered company Springer Nature Switzerland AG
The registered company address is: Gewerbestrasse 11, 6330 Cham, Switzerland

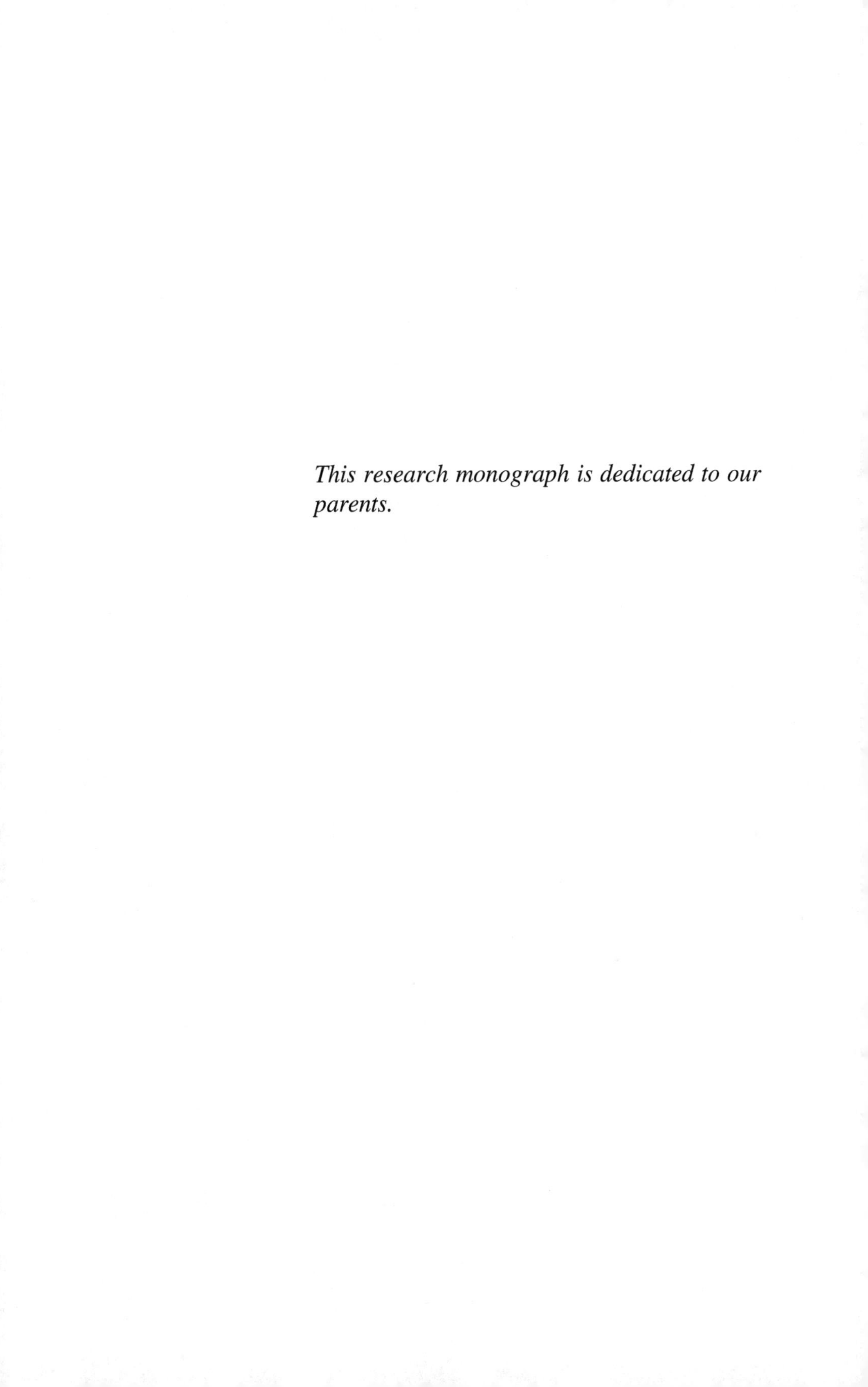

This research monograph is dedicated to our parents.

Preface

The past decades have witnessed an increasing popularity in the Takagi–Sugeno (T–S) fuzzy model since it has powerful capabilities in exactly modeling complicated nonlinear systems via transforming them into a family of local linear subsystems. An overall T–S fuzzy model can be seen as a "combination" of those local linear subsystems with membership functions and IF-THEN rules. Many related works have been reported and the theoretic results have been applied to various fields, such as communication networks, mechanical systems, and power electronic systems. On the other hand, practical systems often experience abrupt variations resulting from stochastic component failures, unexpected environment disturbances, changes between subsystems, and so on. Random switching mechanisms have been used to model these complicated phenomena, such as the Markov jump principle. In the past decades, fuzzy systems with switched parameters have received considerable attention and a large number of results have been published.

This book presents the recent advances in analysis and synthesis of fuzzy systems with switched parameters. Chapter 1 provides an overview of recent developments of fuzzy systems with switched parameters. Chapter 2 investigates the reliable control problem for T–S fuzzy systems with switched actuator failures and quantization. Chapter 3 is concerned with the non-fragile guaranteed cost control problem for T–S fuzzy systems with Markov jump parameters and time-varying delays. Based on the hidden Markov model, Chap. 4 studies the asynchronous quantized control problem for T–S fuzzy Markov jump systems. Chapter 5 focuses on the asynchronous dissipative control problems for both continuous-time and discrete-time systems. By the static output feedback control method, Chap. 6 addresses the extended-dissipative control problem for T–S fuzzy systems with asynchronous modes and intermittent measurements. Chapter 7 considers the asynchronous sliding mode control problem for T–S fuzzy Markov jump systems with matched uncertainties. The H_∞ and L_2–L_∞ filtering problems are discussed in Chap. 8 for T–S fuzzy switched systems with quantization, respectively. Chapters 9 and 10 deal with the reliable filtering problems for T–S fuzzy switched systems in discrete-time and continuous-time domains, respectively. The dissipative asynchronous filter design problem is solved in Chap. 11 for continuous-time T–S fuzzy

Markov jump systems with the hidden Markov model. The networked fault detection problem is investigated in Chap. 12 for fuzzy Markov jump systems.

It is our hope that this book will be a helpful reference for people working in the field of systems and control by fuzzy modeling techniques with switched parameters.

Hangzhou, China Shanling Dong
Hangzhou, China Zheng-Guang Wu
Adelaide/Melbourne, Australia Peng Shi
September 2019

About This Book

This book presents recent advancements on control and filtering design for Takagi–Sugeno (T–S) fuzzy systems with switched parameters. The T–S fuzzy model has received a great deal of attention from people working in the field of control science and engineering since it has powerful ability in transforming complicated nonlinear systems into a set of linear subsystems. Typical applications of T–S fuzzy systems include communication networks, mechanical systems, power electronic systems, and so on. Practical systems often experience abrupt variations in their parameters or structures led by outside disturbance, component failures, and so on. Random switching mechanisms have been used to model these stochastic changes, such as the Markov jump principle. In the past decades, a plenty of results on fuzzy systems with switched parameters have been reported.

In general, there are three kinds of the controller/filter for fuzzy Markov jump systems, namely, the mode-independent controller/filter, the mode-dependent controller/filter, and the asynchronous one. Compared with the mode-dependent case, mode-independence does not focus on whether modes are accessible, ignores partially useful mode information, and thus results in some conservatism. The mode-dependent design approach needs us to timely acquire complete and correct information about the mode from the studied plant. In fact, factors like component failures and data dropouts often make it difficult to obtain exact mode message, which further let the mode-dependent controllers/filters less useful. Recently, to overcome these issues, researchers devote themselves to studying the asynchronous technique. Modes of asynchronous controllers/filters are accessed by observing the original systems based on certain probabilities. In this book, we will investigate the problems of the mode-independent, mode-dependent, and asynchronous controller/filter design.

In our study, some networked constraints, such as data dropouts and time delays, are also considered. Based on Lyapunov function and matrix inequality techniques, performances of the targeted systems are analyzed, including the stochastic stability, dissipativity, H_∞, and so on. This book not only shows how these approaches solve the control and filtering problems effectively, but also gives the potential and meaningful research directions and ideas, which will help the readers understand this field thoroughly.

The book covers many fields including continuous-time and discrete-time Markov processes, fuzzy systems, robust control, and filter design problems. Investigation on these aspects is meaningful both from the theoretical and practical points of view. It is primarily intended for the researchers in system and control theory. It may also function as a valuable reference to lead graduate students and undergraduate students to an active and interesting control research field. The book provides many cases of fuzzy control problems which may be valuable materials for scientists, engineers, and researchers in the field of intelligent control. Also, the contents of this book can be used for advanced courses focusing on fuzzy modeling, analysis, and control.

Contents

Symbols and Acronyms

R	Real number
R^n	n-Dimensional real number vector space
$\Pr\{x\}$	Probability function of x
$\Pr\{x\|y\}$	Conditional probability function of x on y
$E(\cdot)$ or $E\{\cdot\}$	Mathematical expectation
$X > 0 (X \geq 0)$	Matrix X is symmetric positive definite (semi-definite)
$X < 0 (X \leq 0)$	Matrix X is symmetric negative definite (semi-negative)
$\mathrm{tr}(X)$	Trace of matrix X
$\det(X)$	Determinant of matrix X
X^T	Transpose of matrix X
$\mathrm{diag}\{X_1, X_2, \ldots, X_n\}$	Diagonal matrix with X_i as its ith diagonal element
$\begin{bmatrix} A & B \\ * & C \end{bmatrix}$ or $\begin{bmatrix} A & * \\ B & C \end{bmatrix}$	Asterisk $*$ stands for a symmetric matrix B^T in a matrix
sup	Supremum
inf	Infimum
AFISMC	Asynchronous fuzzy integral sliding mode control
GCC	Guaranteed cost control
HMM	Hidden Markov model
LMI	Linear matrix inequality
MJLSs	Markov jump linear systems
MJSs	Markov jump systems
PDC	Parallel distributed compensation
SMC	Sliding mode control
T–S	Takagi–Sugeno

Chapter 1
Introduction

1.1 Fuzzy Systems with Switched Parameters

It is well known that nonlinearity exists widely in practical industrial systems. To deal with it, various efficient approaches have been put forward, including the sliding mode control (SMC) law, the Lipschitz continuity technique and the smoothness approach. The T–S fuzzy model has been recognized as one of powerful and efficient tools in approximating nonlinear systems by dividing the original system into a family of linear subsystems with fuzzy rules and blending all subsystems with membership functions. It has been utilized extensively in many systems, such as networked control systems, manufacturing processes, chemical processes and robotic systems. Considerable attention has been paid to the analysis and synthesis of T–S fuzzy systems, for instance, the filtering design [1–3], the robust control [4–6], the dissipativity issue and the model approximation problem [7]. The T–S fuzzy model has been employed in [8] to investigate the adaptive finite-time stabilization issue for nonlinear systems with uncertain parameters. Via T–S fuzzy knowledge, the work in [9] has accurately modelled the nonlinear stochastic jump diffusion financial system for simplifying the investment policy. The event-triggered control issue has been studied in [10] for networked T–S fuzzy systems.

People have made full use of fuzzy ideas to devise the local controller for the corresponding linearized subsystem, which is known as the parallel distributed compensation (PDC) method. For the PDC approach, fuzzy sets of the designed controller are the same as those of the fuzzy system and they share the identical fuzzy rules. Through the PDC technique, the guaranteed cost controller has been obtained for continuous-time networked systems in [11]. The work in [12] has studied the robust passive fuzzy control issue where constant delays and sampling intervals have been considered. The quantized tracking issue has been investigated in [13] by the PDC and fuzzy adaptive approaches. To analyze the track control problem for sampled-data networked control systems, the work in [14] has adopted the PDC approach to construct an error model. The PDC method has been used in [15] to analyze the

S. Dong et al., *Control and Filtering of Fuzzy Systems with Switched Parameters*, Studies in Systems, Decision and Control 268,
https://doi.org/10.1007/978-3-030-35566-1_1

guaranteed cost control (GCC) issue for bio-economic singular systems. Via using the PDC approach, the hybrid robust boundary and fuzzy control design issue has been studied in [16] for disturbance attenuation of coupled systems.

Owing to abrupt changes in systems, for instance, component failures and environment disturbances, a number of researchers have become interested in introducing switching mechanisms into describing sudden variations of dynamic systems. Thus, dynamic systems with switched parameters are modelled as switched systems. Switched systems [17–19], a special type of hybrid systems, are represented by finite continuous-time or discrete-time subsystems which are under the control of switching rules at each time instant. The work in [20] has adopted sojourn probabilities to describe the random switching situation of discrete-time systems when designing an H_∞ filter. There are two main switching rules. One is the deterministic switching rule, such as the state-dependent switching rule, the time-dependent switching rule and the average dwell-time rule; and the other is the random switching rule, such as the Markov jump rule. The latter rule is mainly investigated in this book.

As a significant stochastic theory, the Markov jump principle has been employed extensively to model random switching phenomena in various aspects, such as communication networks, manufacturing systems and aerospace systems. And modeled systems are called as Markov jump systems (MJSs). In MJSs, each subsystem can be regarded as a mode and whether each subsystem works or not is in the control of transition rates/probabilities [21–24]. The work in [6] has investigated the H_∞ mode-dependent (synchronous) state feedback control problem for nonlinear MJSs with time-varying delays and data dropouts by using the T–S fuzzy approach. The Markov jump theory has been used in [25] to model the generalized neural networks, which consist of finite subsystems. Via adaptive dynamic programming, the work in [26] has investigated the optimal control technique for nonlinear MJSs. The L_2–L_∞ mode-dependent filtering problem has been analyzed in [27] for nonlinear nonhomogeneous MJSs via the T–S fuzzy technique.

In general, there exist three-type approaches in designing a controller/filter for MJSs, namely, mode-independent, mode-dependent and asynchronous methods. Most results in literature on MJSs generally assume that the relationship between the designed controller/filter and the plant is mode-independent or mode-dependent. Mode-independence means that available mode information is never used and the controllers/filters have only one mode while mode-dependent controllers/filters mean that modes of the designed controllers/filters are the same as those of the original systems (mode-dependent controllers/filters are also regarded as synchronous ones). For the control/filtering problem, the mode-dependent approach is highly likely to obtain better performance than the mode-independent one since (i) mode dependence can make full use of the plant mode information; and (ii) mode-independent controllers/filters do not focus on whether modes are accessible and thus ignore partially available mode information. However, the mode-dependent design approach requires us to timely obtain complete and accurate messages about the mode from the studied plant. In practice, factors like component failures, time delays and data dropouts often make us difficult in achieving exact mode information, which further let the mode-dependent controllers/filters ideal. Recently, to overcome these issues, researchers devote themselves to investigating the asynchronous technique and some excellent works have been published. Modes of asynchronous controllers/filters are imprecise since the modes are accessed by observing the original systems based on certain probabilities, which can not ensure that the observed mode is the same as that of original systems. The piecewise homogenous Markov jump has been used in designing the asynchronous filter in [28–30] via utilizing the original system mode and the filter previous mode. There is another non-synchronous mechanism introduced, that is the hidden Markov Model (HMM), which only depends on the original system modes. This principle has been employed in [31] to represent the non-synchronous mode between the controller and the plant.

In control systems, the estimation problem has greatly inspired the interest of researchers from various fields since the information of states is not always accessible and the state estimator can be employed to reconstruct the unmeasurable states and filter the external noise. Kalman filtering [32], one of the well-known approaches, has shown excellent potential in minimizing the variance of the estimation error. However, when there is no sufficient information of the external noise, it no longer produces satisfactory performances. To deal with this problem, some alternative approaches are introduced, such as H_∞ filtering and L_2–L_∞ filtering. The objective of H_∞ filtering is to minimize the energy of the estimation error for the worst possible bounded energy disturbance. It has been recognized to be one of the most appropriate methods in solving the external noise problem with unknown statistics [33–35]. The main purpose of L_2–L_∞ filtering [36, 37] is to minimize the peak value of the estimation error when external noise is bounded energy. The work in [38] has studied the design of reduced-order H_∞ and L_2–L_∞ filtering via linear matrix inequalities (LMIs). The problems of delay-dependent robust H_∞ and L_2–L_∞ filtering have been analyzed in [39]. The work in [40] has investigated the control and filtering problems for discrete linear repetitive processes with H_∞ and L_2–L_∞ performance, respectively.

Dissipativity is another popular research frontier based on the input-output energy consideration, introduced in [41, 42]. It relates the generalized energy supply function with internally stored energy, where the supplied energy is from the exterior environment and the Lyapunov function is analogous to the energy storage function. Dissipative systems only dissipate energy rather than produce it. One important feature of dissipative systems is that the stored energy is less than the supplied energy. Dissipativity contains various basic theories including the circle criterion, the bounded real lemma and the passivity theorem. Moreover, dissipative control can be viewed as a unified framework on robust control problems, for example, H_∞ and passivity control criteria are obtained by adjusting dissipative parameters. Hence, it has wide applications in many fields, for instance, chemical processes and electromechanical systems. And fruitful research results have been obtained as well. The work in [43] has intensively investigated the connections among dissipativity, passivity and positive realness for continuous and discrete time-invariant systems, respectively. Dissipative filtering and control issues have been studied for the two-dimensional Roesser model in [44]. By the dissipative and passive knowledge, the work in [45] has been concerned with investigating stability conditions for large-scale systems.

The work in [46, 47] has analyzed the dissipativity issue for neural networks with time delays. The dissipative control problem about time-varying sampling has been discussed in [48] for T–S fuzzy systems.

Recently, a novel performance index has been introduced, namely, the extended dissipative performance. It is the extension of the common dissipative performance, containing more performance indexes, such as L_2–L_∞ performance, H_∞ performance and passivity. The extended dissipative performance has been used to design the asynchronous and resilient filter for MJSs in [49]. The work in [50] has dealt with the extended dissipative state estimation problem for neural MJSs. The extended dissipative analysis has been studied in [51] for neural networks with time-varying delays.

Besides, in actual implementation, control systems like networked control systems contain a large number of components. In fact, components unavoidably generate unexpected failures which will bring forth some adverse effects on systems, such as performance degradation and even instability. Thus, it deserves our attention to design a reliable controller/filter, which not only operates successfully without failures, but also can tolerate some admissible failures when they happen. Many researchers have done a good job in this subject. The work in [52] has investigated probabilistic sensor failures with a probability density function for discrete time-varying systems. The work in [53] has dealt with the stochastic link failures of sensor networks, which are resulted from probabilistic communication failures and missing measurements. The dissipative reliable filtering issue has been considered in [54] for discrete-time T–S fuzzy systems with time-varying delays. The work in [55] has adopted robust pole region assignment techniques and a pre-compensator to improve the dynamic characteristics against actuator failures.

It is well known that the Lyapunov function plays an important role in analyzing the stability, the stochastic stability and other characteristics of dynamic systems. In dealing with stability and stabilization problems for T–S fuzzy systems, many Lyapunov functions have been developed for reducing conservatism instead of a common Lyapunov function, like the piecewise quadratic Lyapunov function [56] and the fuzzy Lyapunov function [57, 58]. The latter has received extensive investigation in both discrete-time and continuous-time domains. Compared with the discrete-time case, the continuous-time fuzzy Lyapunov function is more challenging since it involves the time derivative of fuzzy membership functions. When investigating the asynchronous dissipative control problem in [59] for the discrete-time fuzzy MJSs, the used Lyapunov function V_k is not only fuzzy-basis-dependent, but also mode-dependent, which means that it can flexibly change with jump modes and fuzzy rules, and bring out less conservatism in spite of computational increase. On the other hand, a non-monotonic Lyapunov function approach has been investigated in [60] for uncertain T–S fuzzy systems, which deserves our further study.

1.2 Book Organization

So far, extensive works have been published on T–S fuzzy systems with switched parameters. However, there lacks a monograph to provide up-to-date developments on this important topic. Thus, the main objective of this book is to fill such gap by providing recent advances in the analysis and synthesis for T–S fuzzy systems with switched parameters. The materials used in the book are mainly based on the research results of the authors.

This book consists of twelve chapters.

Chapter 1 introduces recent developments of fuzzy systems with switched parameters.

Chapter 2 is concerned with the problem of reliable switched controller design for a class of discrete-time T–S fuzzy systems with randomly occurring infinite-distributed delays, quantization and actuator failures. A random Bernoulli process is used to describe the stochastic infinite-distributed delays. Due to limited communication capacity, the control signal is quantized before being transmitted to the actuator by the logarithmic quantizer. We apply the switching mechanism to categorize the stochastic behavior of actuator faults. Based on the PDC method, the switched feedback controller is designed. By the fuzzy-basis-dependent Laypunov functional approach, sufficient conditions are obtained to ensure that the resulting closed-loop system is exponentially stable in the mean-square sense with a given L_2–L_∞ performance index.

Chapter 3 investigates the problem of non-fragile GCC for discrete-time T–S fuzzy MJSs with time-varying delays. With the help of the PDC approach, a non-fragile fuzzy controller is designed. Then via the Lyapunov–Krasovskii function approach, sufficient conditions are obtained ensuring that the resulting closed-loop system is stochastically stable with an upper bound of the guaranteed cost index. The optimal upper bound of the guaranteed cost index and the controller gain can be achieved via the optimization technique.

Chapter 4 considers the problem of asynchronous GCC for fuzzy MJSs with stochastic quantization. The HMM theory is used to describe the non-synchronous controller and the random quantization phenomenon. Based on the T–S fuzzy technique and the Lyapunov function approach, sufficient conditions are obtained, which can not only ensure the stochastic stability of the closed-loop system and the existence of the desired controller, but also can yield the minimal upper bound of GCC performance.

Chapter 5 addresses the strictly dissipative control problem for MJSs via T–S fuzzy rules in continuous-time and discrete-time domains, respectively. The modes of devised fuzzy controller are assumed to run asynchronously with the modes of original systems, which is widespread in practice and described through an HMM. Sufficient conditions are acquired to ensure the stochastic stability and the strict dissipativity of the closed-loop systems, based on which the design methods of the asynchronous fuzzy controller are provided.

Chapter 6 focuses on the problem of asynchronous output feedback control for T–S fuzzy switched systems subject to intermittent measurements. Bernoulli process is employed to model the phenomenon of stochastic intermittent measurements. Based on the HMM and output measurements, an asynchronous controller is designed. Then, sufficient conditions for the existence of an asynchronous controller are proposed, which ensure the stochastic stability of the closed-loop system with a desired extended dissipative performance.

Chapter 7 considers the problem of dissipativity-based asynchronous fuzzy integral SMC for T–S fuzzy MJSs, which are subject to external noises and matched uncertainties. Since modes of original systems cannot be directly obtained, an HMM is employed to detect mode information. With the detected mode and the PDC approach, a suitable fuzzy integral sliding surface is devised. Then using the Lyapunov function, a sufficient condition for the existence of sliding mode controller gains is developed, which can also ensure the stochastic stability of the sliding mode dynamics with a satisfactory dissipative performance. An asynchronous fuzzy integral sliding mode control law is proposed to drive system trajectories into the predetermined sliding mode boundary layer in finite time. For the case with unknown bound of uncertainties, an adaptive asynchronous fuzzy integral sliding mode control (AFISMC) law is developed as well. The studied T–S fuzzy MJSs involve both continuous-time and discrete-time domains.

Chapter 8 studies H_∞ and L_2–L_∞ filtering design problems for discrete-time fuzzy switched systems with quantized measurements. The systems under consideration inherently combine features of the switched hybrid systems and the T–S fuzzy systems. The sector bound approach is employed to deal with quantization effects. Based on the fuzzy-basis-dependent Lyapunov function, sufficient conditions are established such that the filtering error system is stochastically stable and a prescribed noise attenuation level in an H_∞ or L_2–L_∞ sense is achieved.

Chapter 9 deals with the problem of asynchronous and reliable filter design with the performance constraint for fuzzy MJSs. The nonstationary Markov chain is adopted to represent the asynchronous situation between the designed filter and the considered system. By using the mode-dependent Lyapunov function approach and the relaxation matrix technique, a sufficient condition is proposed to ensure that the filtering error system, which is a dual randomly switched system, is stochastically stable and satisfies a given L_2–L_∞ performance index simultaneously. Two different approaches are developed to construct the asynchronous and reliable filter. Owing to Finsler's lemma, the second approach has fewer decision variables and less conservatism than the first one.

Chapter 10 analyzes the reliable L_2–L_∞ filter design problem for nonlinear continuous-time MJSs based on the T–S fuzzy model. A stochastic variable is introduced to describe the encountered sensor failures, the value of which is dependent on the considered plant mode based on a hidden Markov process. In practice, it is general that the information on plant modes is not fully accessible to the reliable filter, which results in the non-synchronous phenomena between the modes of the involved plant and the filter, and has negative effect on the system performance. An HMM is also adopted to depict such kind of non-synchronous phenomena. The filtering error systems go by the name of fuzzy dual hidden MJSs correspondingly. A sufficient condition is proposed for the filtering error systems to ensure the stochastic stability and the guaranteed L_2–L_∞ performance. And the explicit design method of non-synchronous filter gains is given as well.

Chapter 11 addresses the dissipative asynchronous filtering problem for T–S fuzzy MJSs in the continuous-time domain. An HMM is applied to describe the asynchronous situation between the designed filter and the original system. Based on the stochastic Lyapunov function, a sufficient condition is developed to guarantee the stochastic stability of the filtering error systems with a given dissipative performance. Two different methods for the existence of the desired filter are established.

Chapter 12 deals with the problem of dissipativity-based asynchronous fault detection for T–S fuzzy MJSs with networked data dropouts. It is assumed that data dropouts happen intermittently from the plant to the fault detection filter, which is described by Bernoulli process. The HMM is employed to describe the asynchronous phenomenon between the plant and the filter. Based on the Lyapunov theory, a sufficient condition is developed to guarantee that the fault detection system is stochastically stable with strictly dissipative performance. By choosing an appropriate Lyapunov function with the slack matrix technique and Finsler's Lemma, two approaches are proposed to compute filter gains in the form of LMIs.

References

1. Chang, X.-H., Park, J.H., Shi, P.: Fuzzy resilient energy-to-peak filtering for continuous-time nonlinear systems. IEEE Trans. Fuzzy Syst. **25**(6), 1576–1588 (2017)
2. Zhang, S., Wang, Z., Ding, D., Shu, H.: Fuzzy filtering with randomly occurring parameter uncertainties, interval delays, and channel fadings. IEEE Trans. Cybern. **44**(3), 406–417 (2014)
3. Luo, Y., Wang, Z., Wei, G., Alsaadi, F.E.: Robust H_∞ filtering for a class of two-dimensional uncertain fuzzy systems with randomly occurring mixed delays. IEEE Trans. Fuzzy Syst. **25**(1), 70–83 (2017)
4. Dong, S., Wu, Z.-G., Shi, P., Su, H., Lu, R.: Reliable control of fuzzy systems with quantization and switched actuator failures. IEEE Trans. Syst., Man, Cybern.: Syst. **47**(8), 2198–2208 (2017)
5. Wu, Z.-G., Dong, S., Shi, P., Su, H., Huang, T., Lu, R.: Fuzzy-model-based nonfragile guaranteed cost control of nonlinear Markov jump systems. IEEE Trans. Syst., Man, Cybern.: Syst. **47**(8), 2388–2397 (2017)
6. Zhang, L., Ning, Z., Shi, P.: Input-output approach to control for fuzzy Markov jump systems with time-varying delays and uncertain packet dropout rate. IEEE Trans. Cybern. **45**(11), 2449–2460 (2015)
7. Su, X., Wu, L., Shi, P.: Model approximation for fuzzy switched systems with stochastic perturbation. IEEE Trans. Fuzzy Syst. **23**(5), 1458–1473 (2015)
8. Li, Y., Liu, L., Feng, G.: Adaptive finite-time controller design for T-S fuzzy systems. IEEE Trans. Cybern. **47**(9), 2425–2436 (2017)
9. Wu, C.-F., Chen, B.-S., Zhang, W.: Multiobjective investment policy for a nonlinear stochastic financial system: A fuzzy approach. IEEE Trans. Fuzzy Syst. **25**(2), 460–474 (2017)
10. Peng, C., Ma, S., Xie, X.: Observer-based non-PDC control for networked T-S fuzzy systems with an event-triggered communication. IEEE Trans. Cybern. **47**(8), 2279–2287 (2017)
11. Lu, R., Cheng, H., Bai, J.: Fuzzy-model-based quantized guaranteed cost control of nonlinear networked systems. IEEE Trans. Fuzzy Syst. **23**(3), 567–575 (2015)

12. Wu, Z.-G., Shi, P., Su, H., Chu, J.: Network-based robust passive control for fuzzy systems with randomly occurring uncertainties. IEEE Trans. Fuzzy Syst. **21**(5), 966–971 (2013)
13. Liu, Z., Wang, F., Zhang, Y., Chen, C.: Fuzzy adaptive quantized control for a class of stochastic nonlinear uncertain systems. IEEE Trans. Cybern. **46**(2), 524–534 (2016)
14. Xiao, H.-Q., He, Y., Wu, M., Xiao, S.-P., She, J.: New results on H_∞ tracking control based on the T-S fuzzy model for sampled-data networked control system. IEEE Trans. Fuzzy Syst. **23**(6), 2439–2448 (2015)
15. Li, L., Zhang, Q., Zhu, B.: Fuzzy stochastic optimal guaranteed cost control of bio-economic singular Markovian jump systems. IEEE Trans. Cybern. **45**(11), 2512–2521 (2015)
16. Feng, S., Wu, H.-N.: Hybrid robust boundary and fuzzy control for disturbance attenuation of nonlinear coupled ODE-beam systems with application to a flexible spacecraft. IEEE Trans. Fuzzy Syst. **25**(5), 1293–1305 (2017)
17. Mathiyalagan, K., Su, H., Shi, P., Sakthivel, R.: Exponential H_∞ filtering for discrete-time switched neural networks with random delays. IEEE Trans. Cybern. **45**(4), 676–687 (2015)
18. Sanatkar, M.R., White, W.N., Natarajan, B., Scoglio, C.M., Garrett, K.A.: Epidemic threshold of an SIS model in dynamic switching networks. IEEE Trans. Syst., Man, Cybern.: Syst. **46**(3), 345–355 (2016)
19. Ding, Z., Zhou, Y., Jiang, M., MengChu, Z.: A new class of petri nets for modeling and property verification of switched stochastic systems. IEEE Trans. Syst., Man, Cybern.: Syst. **45**(7), 1087–1100 (2015)
20. Tian, E., Wong, W.K., Yue, D., Yang, T.-C.: H_∞ filtering for discrete-time switched systems with known sojourn probabilities. IEEE Trans. Autom. Control **60**(9), 2446–2451 (2015)
21. Zhang, M., Shi, P., Liu, Z., Su, H., Ma, L.: Fuzzy model-based asynchronous H_∞ filter design of discrete-time Markov jump systems. J. Frankl. Inst. **354**(18), 8444–8460 (2017)
22. Wu, Z.-G., Shen, Y., Shi, P., Shu, Z., Su, H.: H_∞ control for 2D Markov jump systems in Roesser model. IEEE Trans. Autom. Control **64**(1), 427–432 (2019)
23. Zhang, L., Prieur, C.: Stochastic stability of Markov jump hyperbolic systems with application to traffic flow control. Automatica **86**, 29–37 (2017)
24. Cui, J., Liu, T., Wang, Y.: New stability criteria for a class of Markovian jumping genetic regulatory networks with time-varying delays. Int. J. Innov. Comput., Inf. Control **13**(3), 809–822 (2017)
25. Samidurai, R., Manivannan, R., Ahn, C.K., Karimi, H.R.: New criteria for stability of generalized neural networks including Markov jump parameters and additive time delays. IEEE Trans. Syst., Man, Cybern.: Syst. **48**(4), 485–499 (2018)
26. Zhong, X., He, H., Zhang, H., Wang, Z.: Optimal control for unknown discrete-time nonlinear Markov jump systems using adaptive dynamic programming. IEEE Trans. Neural Netw. Learn. Syst. **25**(12), 2141–2155 (2014)
27. Yin, Y., Shi, P., Liu, F., Teo, K.L., Lim, C.-C.: Robust filtering for nonlinear nonhomogeneous Markov jump systems by fuzzy approximation approach. IEEE Trans. Cybern. **45**(9), 1706–1716 (2015)
28. Wu, Z.-G., Shi, P., Su, H., Lu, R.: Asynchronous l_2-l_∞ filtering for discrete-time stochastic Markov jump systems with randomly occurred sensor nonlinearities. Automatica **50**(5), 180–186 (2014)
29. Zhang, L., Zhu, Y., Shi, P., Zhao, Y.: Resilient asynchronous H_∞ filtering for Markov jump neural networks with unideal measurements and multiplicative noises. IEEE Trans. Cybern. **45**(12), 2840–2852 (2015)
30. Zhu, Y., Zhang, L., Zheng, W.X.: Distributed H_∞ filtering for a class of discrete-time Markov jump Lur'e systems with redundant channels. IEEE Trans. Ind. Electron. **63**(3), 1876–1885 (2016)
31. Wu, Z.-G., Shi, P., Shu, Z., Su, H., Lu, R.: Passivity-based asynchronous control for Markov jump systems. IEEE Trans. Autom. Control **62**(4), 2020–2025 (2017)
32. Kalman, R.E.: A new approach to linear filtering and prediction problems. J. Basic Eng. **82**(1), 35–45 (1960)

33. Lian, J., Mu, C., Shi, P.: Asynchronous H_∞ filtering for switched stochastic systems with time-varying delay. Inf. Sci. **224**, 200–212 (2013)
34. Zhang, L., Dong, X., Qiu, J., Alsaedi, A., Hayat, T.: H_∞ filtering for a class of discrete-time switched fuzzy systems. Nonlinear Anal.: Hybrid Syst. **14**, 74–85 (2014)
35. Zhang, S., Wang, Z., Ding, D., Dong, H., Alsaadi, F.E., Hayat, T.: Nonfragile fuzzy filtering with randomly occurring gain variations and channel fadings. IEEE Trans. Fuzzy Syst. **24**(3), 505–518 (2016)
36. Wu, H.-N., Wang, J.-W., Shi, P.: A delay decomposition approach to L_2-L_∞ filter design for stochastic systems with time-varying delay. Automatica **47**(7), 1482–1488 (2011)
37. Zhang, H., Shi, Y., Wang, J.: On energy-to-peak filtering for nonuniformly sampled nonlinear systems: a Markovian jump system approach. IEEE Trans. Fuzzy Syst. **22**(1), 212–222 (2014)
38. Grigoriadis, K.M., Watson, J.T.: Reduced-order H_∞ and L_2-L_∞ filtering via linear matrix inequalities. IEEE Trans. Aerosp. Electron. Syst. **33**(4), 1326–1338 (1997)
39. Gao, H., Wang, C.: Delay-dependent robust H_∞ and L_2-L_∞ filtering for a class of uncertain nonlinear time-delay systems. IEEE Trans. Autom. Control **48**(9), 1661–1666 (2003)
40. Wu, L., Lam, J., Paszke, W., Gałkowski, K., Rogers, E., Kummert, A.: Control and filtering for discrete linear repetitive processes with H_∞ and L_2-L_∞ performance. Multidimens. Syst. Signal Process. **20**(3), 235–264 (2009)
41. Willems, J.: Dissipative dynamical systems, Part I: general theory. Arch. Rat. Mech. Anal. **45**(5), 321–393 (1972)
42. Willems, J.: Dissipative dynamical systems, Part II: linear systems with quadratic supply rates. Arch. Rat. Mech. Anal. **45**(5), 321–393 (1972)
43. Kottenstette, N., McCourt, M.J., Xia, M., Gupta, V., Antsaklis, P.J.: On relationships among passivity, positive realness, and dissipativity in linear systems. Automatica **50**(4), 1003–1016 (2014)
44. Ahn, C.K., Shi, P., Basin, M.V.: Two-dimensional dissipative control and filtering for Roesser model. IEEE Trans. Autom. Control **60**(7), 1745–1759 (2015)
45. Ghanbari, V., Wu, P., Antsaklis, P.J.: Large-scale dissipative and passive control systems and the role of star and cyclic symmetries. IEEE Trans. Autom. Control **61**(11), 3676–3680 (2016)
46. Wu, Z.-G., Shi, P., Su, H., Chu, J.: Dissipativity analysis for discrete-time stochastic neural networks with time-varying delays. IEEE Trans. Neural Netw. Learn. Syst. **24**(3), 345–355 (2013)
47. Velmurugan, G., Rakkiyappan, R., Vembarasan, V., Cao, J., Alsaedi, A.: Dissipativity and stability analysis of fractional-order complex-valued neural networks with time delay. Neural Netw. **86**, 42–53 (2017)
48. Wu, Z.-G., Shi, P., Su, H., Lu, R.: Dissipativity-based sampled-data fuzzy control design and its application to truck-trailer system. IEEE Trans. Fuzzy Syst. **23**(5), 1669–1679 (2015)
49. Tao, J., Wu, Z.-G., Su, H., Wu, Y., Zhang, D.: Asynchronous and resilient filtering for Markovian jump neural networks subject to extended dissipativity. IEEE Trans. Cybern. **49**(7), 2504–2513 (2019)
50. Shen, H., Zhu, Y., Zhang, L., Park, J.H.: Extended dissipative state estimation for Markov jump neural networks with unreliable links. IEEE Trans. Neural Netw. Learn. Syst. **28**(2), 346–358 (2017)
51. Lee, T.H., Park, M.-J., Park, J.H., Kwon, O.-M., Lee, S.-M.: Extended dissipative analysis for neural networks with time-varying delays. IEEE Trans. Neural Netw. Learn. Syst. **25**(10), 1936–1941 (2014)
52. Dong, H., Wang, Z., Ding, S.X., Gao, H.: Finite-horizon reliable control with randomly occurring uncertainties and nonlinearities subject to output quantization. Automatica **52**, 355–362 (2015)
53. Liu, J., Wu, C., Wang, Z., Wu, L.: Reliable filter design for sensor networks using type-2 fuzzy framework. IEEE Trans. Ind. Inform. **13**(4), 1742–1752 (2017)
54. Su, X., Shi, P., Wu, L., Basin, M.V.: Reliable filtering with strict dissipativity for T-S fuzzy time-delay systems. IEEE Trans. Cybern. **44**(12), 2470–2483 (2014)

55. Zhao, Q., Jiang, J.: Reliable state feedback control system design against actuator failures. Automatica **34**, 1267–1272 (1998)
56. Li, L., Ding, S.X., Qiu, J., Yang, Y., Zhang, Y.: Weighted fuzzy observer-based fault detection approach for discrete-time nonlinear systems via piecewise-fuzzy Lyapunov functions. IEEE Trans. Fuzzy Syst. **24**(6), 1320–1333 (2016)
57. Liu, Y., Wu, F., Ban, X.: Dynamic output feedback control for continuous-time T-S fuzzy systems using fuzzy Lyapunov functions. IEEE Trans. Fuzzy Syst. **25**(5), 1155–1167 (2017)
58. Kim, S.H.: Relaxation technique for a T-S fuzzy control design based on a continuous-time fuzzy weighting-dependent Lyapunov function. IEEE Trans. Fuzzy Syst. **21**(4), 761–766 (2013)
59. Wu, Z.-G., Dong, S., Su, H., Li, C.: Asynchronous dissipative control for fuzzy Markov jump systems. IEEE Trans. Cybern. **48**(8), 2426–2436 (2018)
60. Nasiri, A., Nguang, S.K., Swain, A., Almakhles, D.J.: Reducing conservatism in H_∞ robust state feedback control design of T-S fuzzy systems: A non-monotonic approach. IEEE Trans. Fuzzy Syst. **26**(1), 386–390 (2018)

Chapter 2
Reliable Control of Fuzzy Systems with Quantization and Switched Actuator Failures

2.1 Introduction

Along with increasing applications of digital equipments in control systems, discrete-time delays [1–3] generally exist, which strongly impact on the stability and performance of dynamic systems. Significant efforts have been made. The fuzzy filtering problem for T–S fuzzy systems with randomly occurring interval time-varying delays has been investigated in [4]. Since signals are frequently distributed among plenty of parallel pathways in a certain time interval, the distributed delay problem for discrete-time systems has been a serious issue in analysis and design of control systems. The work in [5] has paid attention to the robust control for fuzzy systems with infinite-distributed delays. The state feedback control problem for discrete-time stochastic systems with distributed delays has been studied in [6]. The filtering problem of systems with uncertain distributed delays has been considered in [7]. The works in [8, 9] have analysed the reliable control problem for discrete-time fuzzy systems with infinite-distributed delays. Furthermore, the discrete-time distributed delays are likely to be sent in stochastic means. Infinite-distributed delays have been discussed as randomly occurring via the Bernoulli process in [10].

Networked control systems have won great popularity owing to its simple installations, lower cost, reliable quality and so on. For constraints of communication bandwidth and storage in a network, primitive information needs quantization before being sent. Accordingly, the quantization effect is a non-negligible issue in networked control systems, which is one reason for the degraded performance and instability. Abundant efforts have been devoted to researching it. Via finite-level static quantizers, the work in [11] has analyzed the stability issue for sampled-data switched systems. A novel quantizer has been devised in [12], which combines merits of both uniform and logarithmic quantizers. The work in [13] has employed the quantization input to deal with the adaptive tracking problem for switched stochastic nonlinear systems with time delays. The quantization effect has been considered in [14] when designing the event-drive controller for NCSs with Markov packet losses.

S. Dong et al., *Control and Filtering of Fuzzy Systems with Switched Parameters*, Studies in Systems, Decision and Control 268,
https://doi.org/10.1007/978-3-030-35566-1_2

In this chapter, our purpose is to design the reliable L_2–L_∞ controller for discrete-time fuzzy systems, which should not only operate successfully against quantization, switched actuator failures and randomly occurring infinite-distributed delays, but also guarantee the exponential mean-square stability with a prescribed L_2–L_∞ performance index. The Bernoulli distribution is introduced to describe the stochastic infinite-distributed delays. We assume that the switching mode of the controller runs synchronously with that of the actuator. The switched feedback controller is designed via combining switching mechanisms with the PDC approach. Sufficient conditions are obtained via the Lyapunov functional method, which can guarantee that the closed-loop system is mean-square stable and meets a prescribed L_2–L_∞ performance criterion.

2.2 Preliminary Analysis

In this chapter, the reliable L_2–L_∞ control problem is investigated for the fuzzy system, shown in Fig. 2.1.

Consider the T–S fuzzy model with randomly occurring infinite-distributed delays, represented by the following IF-THEN rules:
Plant rule i: IF θ_{1k} is ς_{i1}, θ_{2k} is ς_{i2}, ..., and θ_{pk} is ς_{ip}, THEN

$$\begin{cases} x_{k+1} = A_i x_k + \alpha_k A_{di} \sum_{d=1}^{\infty} \mu_d x_{k-d} + B_{1i} u_k + D_{1i} w_k, \\ z_k = C_i x_k + B_{2i} u_k + D_{2i} w_k, \\ x_k = \phi_k, k \in \mathbb{Z}^-, \end{cases} \tag{2.1}$$

where $x_k \in R^n$ is the state; $u_k \in R^m$ is the control input; $w_k \in R^q$ is the external disturbance, which belongs to $l_2[0, +\infty\}$; $z_k \in R^s$ is the controlled output; and ϕ_k is the initial state; $A_i \in R^{n \times n}$, $A_{di} \in R^{n \times n}$, $B_{1i} \in R^{n \times m}$, $B_{2i} \in R^{n \times m}$, $C_i \in R^{s \times n}$, $D_{1i} \in R^{n \times q}$, and $D_{2i} \in R^{s \times q}$ are known matrices with $i = \{1, 2, \ldots, r\}$. r,

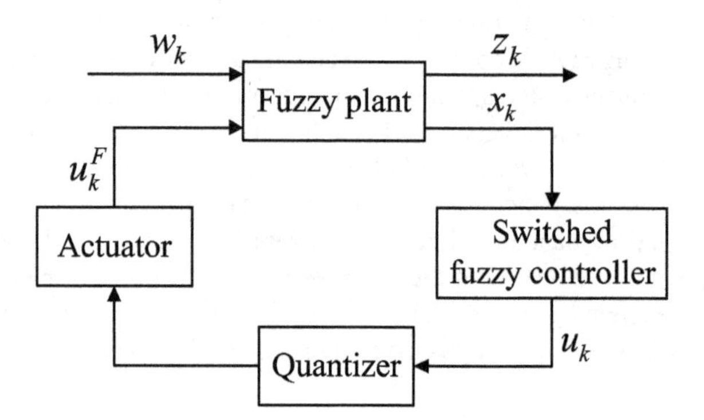

Fig. 2.1 Framework of the closed-loop systems

ς_{ij}, and θ_{jk} ($j = \{1, 2, \ldots, p\}$) are the total quantity of fuzzy rules, the fuzzy set and the premise variable, respectively. $\mathbb{Z}^-$ is the set of non-positive integers. The randomly occurring infinite-distributed delays are described via the stochastic variable α_k, which is subject to the Bernoulli distribution and satisfies

$$\Pr\{\alpha_k = 1\} = \alpha, \ \Pr\{\alpha_k = 0\} = 1 - \alpha. \tag{2.2}$$

It is clear that $\alpha \in [0, 1]$, $E\{\alpha_k\} = \alpha$ and $E\{\alpha_k^2\} = \alpha$ hold.

The constant μ_d ($\mu_d \geq 0$) obeys the following convergence condition:

$$\bar{\mu} = \sum_{d=1}^{\infty} \mu_d < +\infty, \ \sum_{d=1}^{\infty} d\mu_d < +\infty. \tag{2.3}$$

Remark 2.1 In this chapter, we call the term $\sum_{d=1}^{\infty} \mu_d x_{k-d}$ infinite-distributed delays, firstly put forward in [15]. To be more insightful, we had better see it as the discretization of the infinite integral form $\int_{\infty}^{t} K(t-s)x(s)ds$ in continuous systems. Due to abrupt variations of the channels for signal transmissions, time delays occur in a probabilistic way and we describe infinite-distributed delays by the Bernoulli distribution. Moreover, the condition (2.3) is applied to guarantee that the Lyapunov function employed later is convergent.

Utilizing the T–S fuzzy inference method to system (2.1), we obtain the following model:

$$\begin{cases} x_{k+1} = A_k x_k + \alpha_k A_{dk} \sum_{d=1}^{\infty} \mu_d x_{k-d} + B_{1k} u_k + D_{1k} w_k, \\ z_k = C_k x_k + B_{2k} u_k + D_{2k} w_k, \end{cases} \tag{2.4}$$

where

$$A_k = \sum_{i=1}^{r} h_i(\theta_k) A_i, \ A_{dk} = \sum_{i=1}^{r} h_i(\theta_k) A_{di},$$

$$B_{1k} = \sum_{i=1}^{r} h_i(\theta_k) B_{1i}, \ B_{2k} = \sum_{i=1}^{r} h_i(\theta_k) B_{2i},$$

$$C_k = \sum_{i=1}^{r} h_i(\theta_k) C_i, \ D_{1k} = \sum_{i=1}^{r} h_i(\theta_k) D_{1i},$$

$$D_{2k} = \sum_{i=1}^{r} h_i(\theta_k) D_{2i}, \ \theta_k = [\theta_{1k}, \theta_{2k}, \ldots, \theta_{pk}].$$

Define

$$h_i(\theta_k) = \frac{\prod_{j=1}^{p} \varsigma_{ij}(\theta_{jk})}{\sum_{l=1}^{r} \prod_{j=1}^{p} \varsigma_{ij}(\theta_{jk})}, \tag{2.5}$$

where $\varsigma_{ij}(\theta_{jk})$ is the grade of the membership of θ_{jk} in ς_{ij}, and we assume that $\prod_{j=1}^{p} \varsigma_{ij}(\theta_{jk}) \geq 0$. Thus, it is easily observed that $h_i(\theta_k)$ satisfies $h_i(\theta_k) \geq 0$ and $\sum_{i=1}^{r} h_i(\theta_k) = 1$. For convenient notation, we denote $h_i(\theta_k)$ as h_i in the following analysis.

We employ the stochastic variable r_k to describe switching phenomena synchronously happening to the controller and the actuator, and $r_k \in \{1, 2, 3, \ldots, V\}$ satisfies $\Pr\{r_k = v\} = \pi_v$ and $\sum_{v=1}^{V} \pi_v = 1$.

A family of stochastic variables $\bar{r}_v(k)$ are described as

$$\bar{r}_v(k) = \begin{cases} 1, & r_k = v, \\ 0, & r_k \neq v. \end{cases}$$

It follows from the above analysis that

$$E\{\bar{r}_v(k)\} = \pi_v.$$

Remark 2.2 We should be aware that each time, just one subsystem works and they do not interfere with each other. Hence, we have that if $v_1 \neq v_2$, $\Pr\{r_k = v_1, \ r_k = v_2\} = 0$, and

$$\sum_{v=1}^{N} \bar{r}_v(k) = 1, \ E\{\bar{r}_{v_1}(k)\bar{r}_{v_2}(k)\} = \begin{cases} \pi_{v_1}, & v_1 = v_2, \\ 0, & v_1 \neq v_2, \end{cases}$$

where $v_1, v_2 \in \{1, 2, 3, \ldots, V\}$.

Based on the PDC approach, the state feedback controller of system (2.1) is obtained, as shown below.

Controller rule i: IF θ_{1k} is ς_{i1}, θ_{2k} is ς_{i2}, ..., and θ_{pk} is ς_{ip}, THEN

$$u_k = -K_{iv} x_k, \tag{2.6}$$

where $K_{iv} \in \mathbb{R}^{m \times n}$ is the gain matrix of the vth subfuzzy controller.

In fact, due to the limited capacity communication channels, the control signal u_k has to be quantized before being transmitted to the actuator for reducing the quantity of data sent in the network. We adopt the logarithmic quantizer, defined as

$$q(u_k) = [q_1(u_{1k}), q_2(u_{2k}), \ldots, q_m(u_{mk})]^T, \tag{2.7}$$

where m is the number of quantizers, and $q_i(-u_{ik}) = -q_i(u_{ik})$. We employ the set of quantization levels to represent the logarithmic quantizer as follows:

$$Q_i = \{\pm\chi_i^j : \chi_i^j = \rho_i^j \chi_{i0}, \ j = \pm 1, \pm 2, \ldots\} \cup \{\pm\chi_{i0}\} \cup \{0\}, \tag{2.8}$$

where the parameter ρ_i $(0 < \rho_i < 1)$ is the quantization density, and χ_{i0} $(\chi_{i0} > 0)$ is the initial state of the ith quantizer. χ_i^j denotes the output of the ith quantizer at quantization level j. The quantizer $q_i(u_{ik})$ is described as

$$q_i(u_{ik}) = \begin{cases} \chi_i^j, & \text{if } \dfrac{1}{1+\delta_i}\chi_i^j \le u_{ik} \le \dfrac{1}{1-\delta_i}\chi_i^j, \\ 0, & \text{if } u_{ik} = 0, \\ -q_i(-u_{ik}), & \text{if } u_{ik} < 0, \end{cases} \tag{2.9}$$

where $\delta_i = \frac{1-\rho_i}{1+\rho_i}$.

The sector bound method [16] is introduced to cope with the quantization errors as follows:

$$q_i(u_{ik}) - u_{ik} = \Delta_{ik}u_{ik}, \ |\Delta_{ik}| \le \delta_i. \tag{2.10}$$

Accordingly, $q(u_k)$ can be inferred as

$$q(u_k) = (I + \Delta_k)u_k, \tag{2.11}$$

where $\Delta_k = \text{diag}\{\Delta_{1k}, \Delta_{2k}, \ldots, \Delta_{mk}\}$.

From discussion above, the measurement of quantized control signal is expressed as

$$u_k = -(I + \Delta_k)K_{kv}x_k, \tag{2.12}$$

where $K_{kv} = \sum_{i=1}^{r} h_i K_{iv}$.

We adopt the following fault model to model the phenomenon of actuator failures:

$$u_k^F = \beta_{r_k}u_k, \ \beta_{r_k} = \text{diag}\{\beta_{1r_k}, \beta_{2r_k}, \ldots, \beta_{mr_k}\}, \tag{2.13}$$

where u_k^F is the control signal from the actuator, and it is sent to the fuzzy system as shown in Fig. 2.1. The variable β_{ir_k} describes the failure level of the ith actuator, satisfying

$$0 \le \underline{\beta}_{ir_k} \le \beta_{ir_k} \le \overline{\beta}_{ir_k} \le 1, \ \ i = \{1, 2, \ldots, m\}. \tag{2.14}$$

When $r_k = v$, we define

$$\begin{aligned} \acute{\beta}_v &= \text{diag}\left\{\frac{\underline{\beta}_{1v} + \overline{\beta}_{1v}}{2}, \frac{\underline{\beta}_{2v} + \overline{\beta}_{2v}}{2}, \ldots, \frac{\underline{\beta}_{mv} + \overline{\beta}_{mv}}{2}\right\}, \\ \grave{\beta}_v &= \text{diag}\left\{\frac{\overline{\beta}_{1v} - \underline{\beta}_{1v}}{2}, \frac{\overline{\beta}_{2v} - \underline{\beta}_{2v}}{2}, \ldots, \frac{\overline{\beta}_{mv} - \underline{\beta}_{mv}}{2}\right\}, \end{aligned} \tag{2.15}$$

and β_v is rewritten as

$$\beta_v = \acute{\beta}_v + \Lambda_v, \tag{2.16}$$

where $\Lambda_v = \text{diag}\{\Lambda_{1v}, \Lambda_{2v}, \ldots, \Lambda_{mv}\}$ and

$$|\Lambda_{iv}| \leq \frac{\overline{\beta}_{iv} - \underline{\beta}_{iv}}{2}, \quad i = \{1, 2, \ldots, m\}. \tag{2.17}$$

Remark 2.3 Note that β_{iv} is the admissible malfunction level about the ith actuator when the vth failure subsystem is active. If we do not consider the switching phenomenon, the actuator failure (2.13) will become the model in [9]. The outage for the ith actuator in the vth failure state will happen when $\underline{\beta}_{iv} = \overline{\beta}_{iv} = 0$. If $\underline{\beta}_{iv} \neq 0$ and $\overline{\beta}_{iv} \neq 1$, it means that the ith actuator will lead to partial failures in the vth failure state. If $\underline{\beta}_{iv} = \overline{\beta}_{iv} = 1$, the ith actuator will work successfully without failures.

By substituting u_k^F for u_k in (2.4) and considering (2.12), the closed-loop fuzzy system is inferred as

$$\begin{cases} x_{k+1} = \bar{A}_{kv} x_k + \alpha_k A_{dk} \sum_{d=1}^{\infty} \mu_d x_{k-d} + D_{1k} w_k, \\ z_k = \bar{C}_{kv} x_k + D_{2k} w_k, \end{cases} \tag{2.18}$$

where

$$\bar{A}_{kv} = A_k - B_{1k}\beta_v(I + \Delta_k)K_{kv} = \sum_{i=1}^{r}\sum_{j=1}^{r} h_i h_j \bar{A}_{ijv},$$

$$\bar{C}_{kv} = C_k - B_{2k}\beta_v(I + \Delta_k)K_{kv} = \sum_{i=1}^{r}\sum_{j=1}^{r} h_i h_j \bar{C}_{ijv},$$

$$\bar{A}_{ijv} = A_i - B_{1i}\beta_v(I + \Delta_k)K_{jv}, \; \bar{C}_{ijv} = C_i - B_{2i}\beta_v(I + \Delta_k)K_{jv}.$$

The following definition and lemmas are introduced, which are essential for the discussion of the results in this chapter.

Definition 2.1 ([5]) For every initial condition ϕ_k, the closed-loop fuzzy system (2.18) with $w_k \equiv 0$ is said to be exponentially stable in the mean-square sense, if there exist scalars α ($\alpha > 0$) and β ($0 < \beta < 1$) such that

$$E[||x_k||^2] < \alpha\beta^k \sup_{s \in \mathbb{Z}^-} ||\phi_s||^2. \tag{2.19}$$

Lemma 2.4 (Schur Complement) *Given constant matrices S_{11}, S_{22} where $S_{11} = S_{11}^{\mathrm{T}}$ and $S_{22} = S_{22}^{\mathrm{T}} > 0$. Then $S_{11} + S_{12}S_{22}^{-1}S_{12}^{\mathrm{T}} < 0$ if and only if*

$$\begin{bmatrix} S_{11} & S_{12} \\ * & -S_{22} \end{bmatrix} < 0 \quad or \quad \begin{bmatrix} -S_{22} & S_{12}^{\mathrm{T}} \\ * & S_{11} \end{bmatrix} < 0. \tag{2.20}$$

Lemma 2.5 ([15]) *For any matrix $M > 0$, vector $x_i \in R^n$, and scalar constants α_i $(i = 1, 2, 3, \ldots)$, if the series concerned are convergent, the following inequality holds:*

$$\left(\sum_{i=1}^{\infty} \alpha_i x_i\right)^T M \left(\sum_{i=1}^{\infty} \alpha_i x_i\right) \leq \left(\sum_{i=1}^{\infty} \alpha_i\right) \sum_{i=1}^{\infty} \alpha_i x_i^T M x_i. \tag{2.21}$$

The goal of the chapter is to design a controller in the form of (2.6) for system (2.1), which should meet the following two requirements simultaneously:
(1) The closed-loop fuzzy system (2.18) is exponentially stable in the mean-square sense when $w_k \equiv 0$;
(2) The closed-loop fuzzy system (2.18) has a given level γ $(\gamma > 0)$ of L_2–L_∞ noise attenuation, namely, under the zero-initial condition for any $w_k \in l_2[0, \infty)$, z_k satisfies

$$\sup_k \sqrt{E[||z_k||^2]} < \gamma \sqrt{\sum_{k=0}^{\infty} ||w_k||^2}. \tag{2.22}$$

2.3 Main Results

In this section, a sufficient condition is firstly given for system (2.18) such that it can be exponentially stable and satisfy a prescribed L_2–L_∞ performance index in the stochastic setup. Then, we develop the controller design method in the form of (2.6) for the closed-loop fuzzy system (2.18).

Theorem 2.6 *The closed-loop fuzzy system (2.18) is exponentially stable in the mean-square sense with a given L_2–L_∞ performance index γ $(\gamma > 0)$, when there exist matrices $P_i > 0$, $Q > 0$, $R_{1iv} > 0$, $R_{2iv} > 0$ and $R_{3iv} > 0$ for any $i, j, t = \{1, 2, \ldots, r\}$ and $v = \{1, 2, \ldots, V\}$ satisfying*

$$\sum_{v=1}^{V} \pi_v \begin{bmatrix} R_{1iv} & 0 \\ * & R_{2iv} \end{bmatrix} < \begin{bmatrix} P_i - \bar{\mu} Q & 0 \\ * & \bar{\mu}^{-1} Q \end{bmatrix}, \tag{2.23}$$

$$\sum_{v=1}^{V} \pi_v R_{3iv} < P_i, \tag{2.24}$$

$$\Sigma_{iitv} < 0, \tag{2.25}$$

$$\Sigma_{ijtv} + \Sigma_{jitv} < 0, \ i < j, \tag{2.26}$$

$$\Phi_{iiv} < 0, \tag{2.27}$$

$$\Phi_{ijv} + \Phi_{jiv} < 0,\ i < j, \tag{2.28}$$

where

$$\Sigma_{ijtv} = \begin{bmatrix} -P_t & 0 & P_t\bar{A}_{ijv} & \alpha P_t A_{di} & P_t D_{1i} \\ * & -P_t & 0 & f P_t A_{di} & 0 \\ * & * & -R_{1iv} & 0 & 0 \\ * & * & * & -R_{2iv} & 0 \\ & * & * & * & -I \end{bmatrix},$$

$$\Phi_{ijv} = \begin{bmatrix} -\gamma^2 I & \bar{C}_{ijv} & D_{2i} \\ * & -R_{3iv} & 0 \\ * & * & -I \end{bmatrix},\ f = \sqrt{\alpha(1-\alpha)}.$$

Proof Firstly, we assume that there exist matrices $P_i > 0$, $R_{1iv} > 0$, $R_{2iv} > 0$ and $R_{3iv} > 0$ for $i,\ j,\ t = \{1, 2, \ldots, r\}$ and $v = \{1, 2, \ldots, V\}$ satisfying (2.23)–(2.28). These matrices are employed to define the following functions:

$$P_k = \sum_{i=1}^{r} h_i P_i,\ P_{k+1} = \sum_{t=1}^{r} h_t^+ P_t,\ R_{1kv} = \sum_{i=1}^{r} h_i R_{1iv},$$
$$R_{2kv} = \sum_{i=1}^{r} h_i R_{2iv},\ R_{3kv} = \sum_{i=1}^{r} h_i R_{3iv},\ h_t^+ = h_t(\theta_{k+1}).$$

Then, through (2.23) and (2.24), we can easily obtain that

$$\sum_{v=1}^{V} \pi_v \begin{bmatrix} R_{1kv} & 0 \\ * & R_{2kv} \end{bmatrix} < \begin{bmatrix} P_k - \bar{\mu} Q & 0 \\ * & \bar{\mu}^{-1} Q \end{bmatrix}, \tag{2.29}$$

and

$$\sum_{v=1}^{V} \pi_v R_{3kv} < P_k. \tag{2.30}$$

Since

$$\begin{aligned} G_1 &= \sum_{t=1}^{r}\sum_{i=1}^{r}\sum_{j=1}^{r} h_t^+ h_i h_j \Sigma_{ijtv} \\ &= \sum_{t=1}^{r} h_t^+ \left(\sum_{i=1}^{r} h_i^2 \Sigma_{iitv} + \sum_{i=1}^{r-1}\sum_{j=i+1}^{r} h_i h_j (\Sigma_{ijtv} + \Sigma_{jitv}) \right), \end{aligned} \tag{2.31}$$

we have the following inequality from (2.25) and (2.26):

$$G_1 < 0, \tag{2.32}$$

where

$$G_1 = \begin{bmatrix} -P_{k+1} & 0 & P_{k+1}\bar{A}_{kv} & \alpha P_{k+1}A_{dk} & P_{k+1}D_{1k} \\ * & -P_{k+1} & 0 & f P_{k+1}A_{dk} & 0 \\ * & * & -R_{1kv} & 0 & 0 \\ * & * & * & -R_{2kv} & 0 \\ & * & * & * & -I \end{bmatrix}. \tag{2.33}$$

Applying the same way to (2.27) and (2.28), we have

$$G_2 < 0, \tag{2.34}$$

where

$$G_2 = \begin{bmatrix} -\gamma^2 I & \bar{C}_{kv} & D_{2k} \\ * & -R_{3kv} & 0 \\ * & * & -I \end{bmatrix}.$$

The following inequalities are obtained by applying Lemma 2.4 to (2.32) and (2.34):

$$G_1' < 0, \tag{2.35}$$

and

$$G_2' < 0, \tag{2.36}$$

where

$$G_1' = \begin{bmatrix} \bar{A}_{kv}^T & 0 \\ \alpha A_{dk}^T & f A_{dk}^T \\ D_{1k}^T & 0 \end{bmatrix} \begin{bmatrix} P_{k+1} & 0 \\ 0 & P_{k+1} \end{bmatrix} \begin{bmatrix} \bar{A}_{kv}^T & 0 \\ \alpha A_{dk}^T & f A_{dk}^T \\ D_{1k}^T & 0 \end{bmatrix}^T - \begin{bmatrix} R_{1kv} & 0 & 0 \\ * & R_{2kv} & 0 \\ & * & I \end{bmatrix},$$

$$G_2' = \begin{bmatrix} \bar{C}_{kv}^T \\ D_{2k}^T \end{bmatrix} \begin{bmatrix} \bar{C}_{kv}^T \\ D_{2k}^T \end{bmatrix}^T - \gamma^2 \begin{bmatrix} R_{3kv} & 0 \\ * & I \end{bmatrix}.$$

From $\sum_{v=1}^{V} \pi_v = 1$ with (2.29) and (2.30), it is easy to obtain that

$$G_1'' < 0, \tag{2.37}$$

and

$$G_2'' < 0, \tag{2.38}$$

where

$$G_1'' = \sum_{v=1}^{V} \pi_v \left(\begin{bmatrix} \bar{A}_{kv}^T & 0 \\ \alpha A_{dk}^T & f A_{dk}^T \\ D_{1k}^T & 0 \end{bmatrix} \begin{bmatrix} P_{k+1} & 0 \\ 0 & P_{k+1} \end{bmatrix} \begin{bmatrix} \bar{A}_{kv}^T & 0 \\ \alpha A_{dk}^T & f A_{dk}^T \\ D_{1k}^T & 0 \end{bmatrix}^T \right)$$
$$- \begin{bmatrix} P_k - \bar{\mu} Q & 0 & 0 \\ * & \bar{\mu}^{-1} Q & 0 \\ * & * & I \end{bmatrix},$$
$$G_2'' = \sum_{v=1}^{V} \pi_v \begin{bmatrix} \bar{C}_{kv}^T \\ D_{2k}^T \end{bmatrix} \begin{bmatrix} \bar{C}_{kv}^T \\ D_{2k}^T \end{bmatrix}^T - \gamma^2 \begin{bmatrix} P_k & 0 \\ * & I \end{bmatrix}.$$

Define the Lyapunov function as

$$V_k = x_k^T P_k x_k + \sum_{d=1}^{\infty} \mu_d \sum_{\tau=k-d}^{k-1} x_\tau^T Q x_\tau. \tag{2.39}$$

Calculating the difference of V_k, we have

$$\begin{aligned} \Delta V_k &= V_{k+1} - V_k \\ &= x_{k+1}^T P_{k+1} x_{k+1} - x_k^T (P_k - \bar{\mu} Q) x_k - \sum_{d=1}^{\infty} \mu_d x_{k-d}^T Q x_{k-d}. \end{aligned} \tag{2.40}$$

From Lemma 2.5, it follows that

$$\sum_{d=1}^{\infty} \mu_d x_{k-d}^T Q x_{k-d} \geq \bar{\mu}^{-1} \left(\sum_{d=1}^{\infty} \mu_d x_{k-d} \right)^T Q \left(\sum_{d=1}^{\infty} \mu_d x_{k-d} \right). \tag{2.41}$$

Note that

$$E\{\bar{r}_{v_1}(k) \bar{r}_{v_2}(k)\} = \begin{cases} \pi_{v_1}, & v_1 = v_2, \\ 0, & v_1 \neq v_2, \end{cases}$$

and $E\{a_k^2\} = a = f^2 + a^2$. It follows from (2.40) and (2.41) along the trajectory of system (2.18) that

$$E\{\Delta V_k - w_k^T w_k\} \leq \zeta_k^T G_1'' \zeta_k, \tag{2.42}$$

where

$$\zeta_k = \left[x_k^T \ \sum_{d=1}^{\infty} \mu_d x_{k-d}^T \ w_k^T \right]^T.$$

It easily yields from (2.37) that $E\{\Delta V_k - w_k^T w_k\} < 0$ holds.

When $w_k \equiv 0$,

$$E\{\Delta V_k\} \le \bar{\zeta}_k^T G_1''' \bar{\zeta}_k < 0, \tag{2.43}$$

where

$$\bar{\zeta}_k = \begin{bmatrix} x_k^T & \sum_{d=1}^{\infty} \mu_d x_{k-d}^T \end{bmatrix}^T,$$

$$G_1''' = \sum_{v=1}^{V} \pi_v \begin{bmatrix} \bar{A}_{kv}^T & 0 \\ \alpha A_{dk}^T & f A_{dk}^T \end{bmatrix} \begin{bmatrix} P_{k+1} & 0 \\ 0 & P_{k+1} \end{bmatrix} \begin{bmatrix} \bar{A}_{kv}^T & 0 \\ \alpha A_{dk}^T & f A_{dk}^T \end{bmatrix}^T - \begin{bmatrix} P_k - \bar{\mu} Q & 0 \\ * & \bar{\mu}^{-1} Q \end{bmatrix}.$$

Thus, there must exist a scalar λ such that $E\{\Delta V_k\} < -\lambda ||x_k||^2$ holds. Through the similar proof used in [17], it is concluded that system (2.18) can keep exponentially stable in the mean-square sense.

Next, we are about to verify that system (2.18) meets the L_2–L_∞ performance criterion under the zero initial state condition.

Consider the following index:

$$J = \sum_{i=0}^{k-1} E\{\Delta V_i - w_i^T w_i\}. \tag{2.44}$$

Due to $E\{\Delta V_k - w_k^T w_k\} < 0$, $J < 0$ holds. Thus, we have

$$J = E\left\{ V_k - \sum_{i=0}^{k-1} w_i^T w_i \right\} < 0. \tag{2.45}$$

It can be clearly observed that

$$x_k^T P_k x_k \le E\{V_k\} < E\left\{ \sum_{i=0}^{k-1} w_i^T w_i \right\}. \tag{2.46}$$

Since

$$E\left\{z_k^T z_k\right\} = \begin{bmatrix} x_k \\ w_k \end{bmatrix}^T \sum_{v=1}^{V} \pi_v \begin{bmatrix} \bar{C}_{kv}^T \\ D_{2k}^T \end{bmatrix} \begin{bmatrix} \bar{C}_{kv}^T \\ D_{2k}^T \end{bmatrix}^T \begin{bmatrix} x_k \\ w_k \end{bmatrix}, \tag{2.47}$$

it implies from (2.38) that

$$E\left\{z_k^T z_k\right\} \le E\{\gamma^2 x_k^T P_k x_k + \gamma^2 w_k^T w_k\}. \tag{2.48}$$

From (2.46), we obtain

$$
\begin{aligned}
E\left\{z_k^T z_k\right\} &\leq E\left\{\gamma^2 \sum_{i=0}^{k-1} w_i^T w_i + \gamma^2 w_k^T w_k\right\} \\
&< E\left\{\gamma^2 \sum_{i=0}^{\infty} w_i^T w_i\right\}.
\end{aligned}
\tag{2.49}
$$

Consequently, it is easy to see that the closed-loop fuzzy system (2.18) satisfies the L_2–L_∞ performance criterion under the zero initial state condition. This proof is finished.

Now, we are in a position to solve the designed controller parameter K_{iv} based on Theorem 2.6.

Theorem 2.7 *The closed-loop fuzzy system (2.18) is exponentially stable in the mean-square sense with a given L_2–L_∞ performance index γ ($\gamma > 0$), when there exist matrices $\tilde{P}_i > 0$, $\tilde{Q} > 0$, $\tilde{R}_{1iv} > 0$, $\tilde{R}_{2iv} > 0$, $\tilde{R}_{3iv} > 0$, X, $\tilde{K}_{iv}$, diagonal matrices $M_{1v} > 0$, $M_{2v} > 0$ and two scalars $\varepsilon_1 > 0$, $\varepsilon_2 > 0$ such that for any $i, j, t = \{1, 2, \ldots, r\}$ and $v = \{1, 2, \ldots, V\}$ satisfying*

$$
\sum_{v=1}^{V} \pi_v \begin{bmatrix} \tilde{R}_{1iv} & 0 \\ * & \tilde{R}_{2iv} \end{bmatrix} < \begin{bmatrix} \tilde{P}_i - \bar{\mu}\tilde{Q} & 0 \\ * & \bar{\mu}^{-1}\tilde{Q} \end{bmatrix},
\tag{2.50}
$$

$$
\sum_{v=1}^{V} \pi_v \tilde{R}_{3iv} < \tilde{P}_i,
\tag{2.51}
$$

$$
\begin{bmatrix}
\tilde{\Sigma}_{iitv}^1 & \Sigma_{iv}^2 & \varepsilon_1 \Sigma_{iv}^3 & \Sigma_{iv}^4 & \Sigma_{iv}^5 M_{1v} \\
* & -\varepsilon_1 I & 0 & 0 & 0 \\
* & * & -\varepsilon_1 I & \varepsilon_1 \Delta & 0 \\
* & * & * & -M_{1v} & 0 \\
* & * & * & * & -M_{1v}
\end{bmatrix} < 0,
\tag{2.52}
$$

$$
\begin{bmatrix}
\Omega_{ijtv}^{11} & \Omega_{ijv}^{12} & \varepsilon_1 \Omega_{ijv}^{13} & \Omega_{ijv}^{14} & \Omega_{ijv}^{15} M_{1v} \\
* & -\varepsilon_1 \Omega^{22} & 0 & 0 & 0 \\
* & * & -\varepsilon_1 \Omega^{33} & \varepsilon_1 \Omega^{34} & 0 \\
* & * & * & -\Omega^{44} & 0 \\
* & * & * & * & -\Omega^{55}
\end{bmatrix} < 0, \ i < j,
\tag{2.53}
$$

$$
\begin{bmatrix}
\tilde{\Phi}_{iiv}^1 & \Phi_{iv}^2 & \varepsilon_2 \Phi_{iv}^3 & \Phi_{iv}^4 & \Phi_{iv}^5 M_{2v} \\
* & -\varepsilon_2 I & 0 & 0 & 0 \\
* & * & -\varepsilon_2 I & \varepsilon_2 \Delta & 0 \\
* & * & * & -M_{2v} & 0 \\
* & * & * & * & -M_{2v}
\end{bmatrix} < 0,
\tag{2.54}
$$

$$\begin{bmatrix} \Psi_{ijtv}^{11} & \Psi_{ijv}^{12} & \varepsilon_2\Psi_{ijv}^{13} & \Psi_{ijv}^{14} & \Psi_{ijv}^{15}M_{2v} \\ * & -\varepsilon_2\Psi^{22} & 0 & 0 & 0 \\ * & * & -\varepsilon_2\Psi^{33} & \varepsilon_2\Psi^{34} & 0 \\ * & * & * & -\Psi^{44} & 0 \\ * & * & * & * & -\Psi^{55} \end{bmatrix} < 0, \ i < j, \tag{2.55}$$

where

$$\tilde{\Sigma}_{ijtv}^{1} = \begin{bmatrix} \Upsilon_t^1 & 0 & A_iX - B_{1i}\acute{\beta}_v\tilde{K}_{jv} & \alpha A_{di}X & D_{1i} \\ * & \Upsilon_t^1 & 0 & fA_{di}X & 0 \\ * & * & -\tilde{R}_{1iv} & 0 & 0 \\ * & * & * & -\tilde{R}_{2iv} & 0 \\ * & * & * & * & -I \end{bmatrix},$$

$$\Sigma_{jv}^{2} = \left[0\ 0\ \tilde{K}_{jv}\ 0\ 0\right]^T, \ \Sigma_{iv}^{3} = \left[-\Delta\acute{\beta}_vB_{1i}^T\ 0\ 0\ 0\ 0\right]^T,$$

$$\Sigma_{jv}^{4} = \left[0\ 0\ \tilde{K}_{jv}\ 0\ 0\right]^T, \ \Sigma_{iv}^{5} = \left[-\grave{\beta}_vB_{1i}^T\ 0\ 0\ 0\ 0\right]^T,$$

$$\tilde{\Phi}_{ijv}^{1} = \begin{bmatrix} -\gamma^2I & \Upsilon_{ijv}^2 & D_{2i} \\ * & -\tilde{R}_{3iv} & 0 \\ * & * & -I \end{bmatrix}, \ \Phi_{jv}^{2} = \left[0\ \tilde{K}_{jv}\ 0\right]^T,$$

$$\Phi_{iv}^{3} = \left[-\Delta\acute{\beta}_vB_{2i}^T\ 0\ 0\right]^T, \ \Phi_{jv}^{4} = \left[0\ \tilde{K}_{jv}\ 0\right]^T,$$

$$\Omega_{ijtv}^{11} = \tilde{\Sigma}_{ijtv}^{1} + \tilde{\Sigma}_{jitv}^{1}, \ \Psi_{ijtv}^{11} = \tilde{\Phi}_{ijtv}^{1} + \tilde{\Phi}_{jitv}^{1},$$

$$\Omega_{ijv}^{12} = \left[\Sigma_{jv}^2\ \Sigma_{iv}^2\right], \ \Psi_{ijv}^{12} = \left[\Phi_{jv}^2\ \Phi_{iv}^2\right],$$

$$\Omega_{ijv}^{13}\left[\Sigma_{iv}^3\ \Sigma_{jv}^3\right], \ \Psi_{ijv}^{13} = \left[\Phi_{iv}^3\ \Phi_{jv}^3\right],$$

$$\Omega_{ijv}^{14} = \left[\Sigma_{jv}^4\ \Sigma_{iv}^4\right], \ \Psi_{ijv}^{14} = \left[\Phi_{jv}^4\ \Phi_{iv}^4\right],$$

$$\Omega_{ijv}^{15} = \left[\Sigma_{iv}^5\ \Sigma_{jv}^5\right], \ \Psi_{ijv}^{15} = \left[\Phi_{iv}^5\ \Phi_{jv}^5\right],$$

$$\Omega^{22} = \Omega^{33} = \Psi^{22} = \Psi^{33} = diag\{I, I\},$$

$$\Omega^{34} = \Psi^{34} = diag\{\Delta, \Delta\}, \ f = \sqrt{\alpha(1-\alpha)}, \ \Phi_{iv}^{5} = \left[-\grave{\beta}_vB_{2i}^T\ 0\ 0\right]^T,$$

$$\Omega^{44} = \Omega^{55} = diag\{M_{1v}, M_{1v}\}, \ \Psi^{44} = \Psi^{55} = diag\{M_{2v}, M_{2v}\},$$

$$\Upsilon_t^1 = \tilde{P}_t - X - X^T, \ \Upsilon_{ijv}^2 = C_iX - B_{2i}\acute{\beta}_v\tilde{K}_{jv}.$$

Furthermore, if inequalities (2.50)–(2.55) have feasible solutions, the controller parameter in the form of (2.6) is given as follows:

$$K_{iv} = \tilde{K}_{iv}X^{-1}. \tag{2.56}$$

Proof Suppose there exist matrices $\tilde{P}_i$, $\tilde{Q}$, $\tilde{R}_{1iv}$, $\tilde{R}_{2iv}$, $\tilde{R}_{3iv}$, X, $\tilde{K}_{iv}$ satisfying (2.50)–(2.55). We define the following functions:

$$\tilde{P}_t = X^T P_{1i} X,\ \tilde{Q} = X^T Q X,\ \tilde{R}_{1iv} = X^T R_{1iv} X,$$
$$\tilde{R}_{2iv} = X^T R_{2iv} X,\ \tilde{R}_{3iv} = X^T R_{3iv} X,\ \tilde{K}_{iv} = K_{iv} X. \tag{2.57}$$

It follows from (2.52) and Lemma 2.4 that

$$\begin{bmatrix} \tilde{\Sigma}^1_{iitv} & \Sigma^2_{iv} & \varepsilon_1 \Sigma^3_{iv} \\ * & -\varepsilon_1 I & 0 \\ * & * & -\varepsilon_1 I \end{bmatrix} + \begin{bmatrix} \Sigma^4_{iv} \\ 0 \\ \varepsilon_1 \Delta \end{bmatrix} M^{-1}_{1v} \begin{bmatrix} \Sigma^4_{iv} \\ 0 \\ \varepsilon_1 \Delta \end{bmatrix}^T + \begin{bmatrix} \Sigma^5_{iv} \\ 0 \\ 0 \end{bmatrix} M_{1v} \begin{bmatrix} \Sigma^5_{iv} \\ 0 \\ 0 \end{bmatrix}^T < 0. \tag{2.58}$$

From (2.15) and (2.17), it yields that

$$\begin{bmatrix} \tilde{\Sigma}^1_{iitv} & \Sigma^2_{iv} & \varepsilon_1 \Sigma^3_{iv} \\ * & -\varepsilon_1 I & 0 \\ * & * & -\varepsilon_1 I \end{bmatrix} + \begin{bmatrix} \Sigma^4_{iv} \\ 0 \\ \varepsilon_1 \Delta \end{bmatrix} M^{-1}_{1v} \begin{bmatrix} \Sigma^4_{iv} \\ 0 \\ \varepsilon_1 \Delta \end{bmatrix}^T + \begin{bmatrix} \Sigma'^5_{iv} \\ 0 \\ 0 \end{bmatrix} M_{1v} \begin{bmatrix} \Sigma'^5_{iv} \\ 0 \\ 0 \end{bmatrix}^T < 0, \tag{2.59}$$

where

$$\Sigma'^5_{iv} = \left[-\Lambda_v B^T_{1i}\ 0\ 0\ 0\ 0\right]^T.$$

Using the elementary inequality $x^T y + y^T x \le \varepsilon x^T x + \varepsilon^{-1} y^T y$, we have

$$\begin{bmatrix} \tilde{\Sigma}^1_{iitv} & \Sigma^2_{iv} & \varepsilon_1 \Sigma^3_{iv} \\ * & -\varepsilon_1 I & 0 \\ * & * & -\varepsilon_1 I \end{bmatrix} + \begin{bmatrix} \Sigma^4_{iv} \\ 0 \\ \varepsilon_1 \Delta \end{bmatrix} \begin{bmatrix} \Sigma'^5_{iv} \\ 0 \\ 0 \end{bmatrix}^T + \begin{bmatrix} \Sigma'^5_{iv} \\ 0 \\ 0 \end{bmatrix} \begin{bmatrix} \Sigma^4_{iv} \\ 0 \\ \varepsilon_1 \Delta \end{bmatrix}^T < 0. \tag{2.60}$$

Via putting another way, the following equation holds:

$$\begin{bmatrix} \tilde{\Sigma}'^1_{iitv} & \Sigma^2_{iv} & \varepsilon_1 \Sigma'^3_{iv} \\ * & -\varepsilon_1 I & 0 \\ * & * & -\varepsilon_1 I \end{bmatrix} < 0, \tag{2.61}$$

where

$$\tilde{\Sigma}'^1_{iitvt} = \begin{bmatrix} \Upsilon^1_t & 0 & A_i X - B_{1i} \beta_v \tilde{K}_{iv} & \alpha A_{di} X & D_{1i} \\ * & \Upsilon^1_t & 0 & f A_{di} X & 0 \\ * & * & -\tilde{R}_{1iv} & 0 & 0 \\ * & * & * & -\tilde{R}_{2iv} & 0 \\ * & * & * & * & -I \end{bmatrix},$$

$$\Sigma'^3_{iv} = \left[-\Delta \beta_v B^T_{1i}\ 0\ 0\ 0\ 0\right]^T.$$

Applying Lemma 2.4 to (2.61), we have

$$\tilde{\Sigma}'^1_{iitv} + \varepsilon_1^{-1} \Sigma^2_{iv} \Sigma^{2T}_{iv} + \varepsilon_1 \Sigma'^3_{iv} \Sigma'^{3T}_{iv} < 0. \tag{2.62}$$

Since Δ and Δ_k are diagonal matrices, satisfying $\Delta_k \Delta^{-1} < I$, we have $(\Delta_k \Delta^{-1})(\Delta_k \Delta^{-1})^T < I$ and $(\Delta_k \Delta^{-1})^T = \Delta^{-1} \Delta_k$. Then it follows that

$$\Sigma''_{iitv} = \tilde{\Sigma}'^{1}_{iitv} + \Sigma^{2}_{iv}\Delta_k\Delta^{-1}\Sigma'^{3T}_{iv} + \Sigma'^{3}_{iv}\Delta^{-1}\Delta_k\Sigma^{2T}_{iv} < 0, \tag{2.63}$$

where

$$\Sigma''_{iitv} = \begin{bmatrix} \Upsilon^1_t & 0 & \Upsilon'_{iiv} & \alpha A_{di}X & D_{1i} \\ * & \Upsilon^1_t & 0 & f A_{di}X & 0 \\ * & * & -\tilde{R}_{1iv} & 0 & 0 \\ * & * & * & -\tilde{R}_{2iv} & 0 \\ * & * & * & * & -I \end{bmatrix},$$

$$\Upsilon'_{iiv} = A_i X - B_{1i}\beta_v(I + \Delta_k)\tilde{K}_{iv}.$$

Because $P_t > 0$ and $\tilde{P} = X^T P_t X$, we have $(P_t^{-1} - X)^T P_t (P_t^{-1} - X) > 0$, and $-P_t^{-1} < \tilde{P}_t - X - X^T$. It is clear to find that

$$\Sigma'''_{iitv} = \begin{bmatrix} -P_t^{-1} & 0 & \Upsilon'_{iiv} & \alpha A_{di}X & D_{1i} \\ * & -P_t^{-1} & 0 & f A_{di}X & 0 \\ * & * & -\tilde{R}_{1iv} & 0 & 0 \\ * & * & * & -\tilde{R}_{2iv} & 0 \\ * & * & * & * & -I \end{bmatrix} < 0. \tag{2.64}$$

On the other hand, pre- and post-multiply (2.25) with $\mathrm{diag}\{P_t^{-1}, P_t^{-1}, I, I, I\}$, and then pre-multiply $\mathrm{diag}\{I, I, X^T, X^T, I\}$ and post-multiply its transpose, respectively, and we have (2.64).

Thus, it is concluded that if (2.52) holds, we will obtain (2.25). We adopt the similar way to test (2.26)–(2.28), (2.53)–(2.55), and the similar results are obtained: when (2.53)–(2.55) are correct, (2.26), (2.27) and (2.28) hold, respectively. Pre-multiplying $\mathrm{diag}\{X^T, X^T\}$ and post-multiplying its transpose to (2.23), we have (2.50). Pre-multiplying X^T and post-multiplying its transpose to (2.24), we have (2.51). This proof is finished.

Remark 2.8 Note that the slack matrix X is employed to eliminate couplings between P_t, $\bar{A}_{ijv}$ and A_{di} in Theorem 2.6. This makes the control design simpler at a certain degree. The number of LMIs and decision variables in Theorem 2.7 are $3Vr + Vr^2$ and $2mV + (0.5 + 0.5r + mVr + 1.5Vr)n + (1.5 + 0.5r + 1.5Vr)n^2$, from which it is easy to observe that computation burden will become heavier with the increase of fuzzy rule r. The local approximation technique for decreasing r has been discussed in [18] and it deserves our more attention.

Remark 2.9 Note that Theorem 2.7 provides a satisfactory solution to the reliable control design problem for fuzzy systems with infinite-distributed delays, the state quantization, actuator failures and the switching phenomenon. The feasibility of the designed controller with the optimal L_2–L_∞ performance can be achieved by coping with the convex optimization problem in the following:

$$\min \quad \sigma \quad \text{subject to (2.50)–(2.55) with } \sigma = \gamma^2. \tag{2.65}$$

2.4 Illustrative Example

In this section, a numerical example is provided to demonstrate the feasibility and effectiveness of the proposed approach in the previous section.

Consider the three-order fuzzy system (2.1) (rule = 2) with the following parameters, which have been studied in [5, 8].

$$\left[A_1 \middle| A_2 \right] = \left[\begin{array}{ccc|ccc} 1 & 0.31 & 0 & 0.8 & -0.38 & 1 \\ 0 & 0.33 & 0.21 & -0.2 & 0 & 0.21 \\ 0 & 0 & -0.52 & 0.1 & 0 & -0.55 \end{array} \right],$$

$$\left[A_{d1} \middle| A_{d2} \right] = \left[\begin{array}{ccc|ccc} 0.2 & 0.1 & 0 & 0 & -0.21 & 0 \\ 0.1 & -0.1 & 0 & 0.31 & 0.1 & 0 \\ 0 & 0.2 & -0.1 & 0 & -0.22 & 0.1 \end{array} \right],$$

$$\left[B_{11} \middle| B_{12} \middle| B_{21} \middle| B_{22} \right] = \left[\begin{array}{cc|cc|cc|cc} 1 & 1 & 1 & 0 & 1 & 1 & 1 & 1 \\ 0 & 1 & 0 & 1 & 0 & 1 & 0 & 1 \\ 0 & 1 & 0 & 1 & 0 & 1 & 0 & 1 \end{array} \right],$$

$$\left[C_1 \middle| C_2 \right] = \left[\begin{array}{ccc|ccc} -0.02 & 0 & 0 & -0.12 & 0 & 0.1 \\ 0 & 0 & 0 & 0 & -0.31 & 0.1 \\ 0 & 0 & -0.1 & 0 & 0.2 & -0.1 \end{array} \right],$$

$$\left[D_{11} \middle| D_{12} \middle| D_{21} \middle| D_{22} \right] = \left[\begin{array}{c|c|c|c} 2.1 & 2 & 2.15 & 2 \\ 0 & 2.12 & 0 & 2 \\ 0 & 2 & 0 & 0.22 \end{array} \right].$$

The normalized fuzzy weighting functions are assumed to be

$$h_1 = \frac{1 - \sin^2(x_{1k})}{2}, \ h_2 = 1 - h_1.$$

The Bernoulli process is assumed to satisfy $\Pr\{a_k = 1\} = 0.8$. Let $\mu_d = 2^{-3-d}$, and we obtain that

$$\bar{\mu} = \frac{1}{8}, \ \sum_{d=1}^{\infty} d\mu_d = \frac{1}{4},$$

which meets the convergence condition. There are two quantizers, whose quantization densities are 0.8182 and 0.7391, respectively. We assume that there are three switching modes with switching probabilities $\pi_1 = 0.4$, $\pi_2 = 0.3$ and $\pi_3 = 0.3$. At the same time, we assume that actuator failures are

$$\left[\begin{array}{c|c} \beta_1 & \beta_2 \\ \hline \beta_3 & \end{array} \right] = \left[\begin{array}{cc|cc} \beta_{11} & 0 & \beta_{12} & 0 \\ 0 & \beta_{21} & 0 & \beta_{22} \\ \hline \beta_{13} & 0 & & \\ 0 & \beta_{23} & & \end{array} \right],$$

where

$$\beta_{11} \in [0.8,\ 1],\ \beta_{12} \in [0.9,\ 1],$$
$$\beta_{21} \in [0.9,\ 1],\ \beta_{22} \in [0.8,\ 0.9],$$
$$\beta_{13} \in [0.8,\ 0.9],\ \beta_{23} \in [0.85,\ 0.95].$$

According to (2.15) and (2.17), one has

$$\left[\begin{array}{c|c|c} \acute{\beta}_1 & \acute{\beta}_2 & \acute{\beta}_3 \\ \hline \grave{\beta}_1 & \grave{\beta}_2 & \grave{\beta}_3 \end{array}\right] = \left[\begin{array}{cc|cc|cc} 0.9 & 0 & 0.95 & 0 & 0.85 & 0 \\ 0 & 0.95 & 0 & 0.85 & 0 & 0.9 \\ \hline 0.1 & 0 & 0.05 & 0 & 0.05 & 0 \\ 0 & 0.05 & 0 & 0.05 & 0 & 0.05 \end{array}\right].$$

By using LMI Toolbox in Matlab, we can solve LMIs (2.50)–(2.55), and achieve the optimal L_2–L_∞ performance $\gamma^* = 2.8434$, and the desired feedback gain matrix as shown in (2.66). When the initial condition is $x_k = [10\ \ -5\ 4]^T$ $(k \in \mathbb{Z}^-)$, Fig. 2.2 gives state trajectories under the noise in Fig. 2.3. This example has demonstrated the correctness of the proposed method.

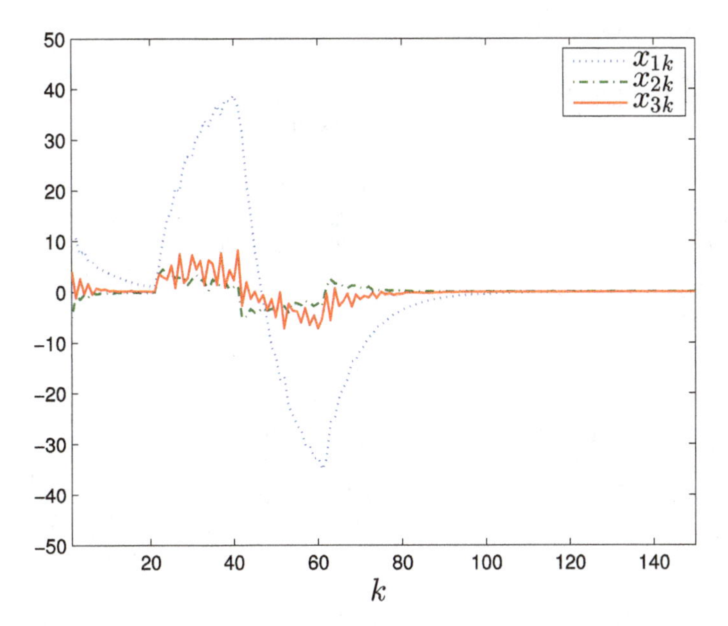

Fig. 2.2 State trajectories of the closed-loop system

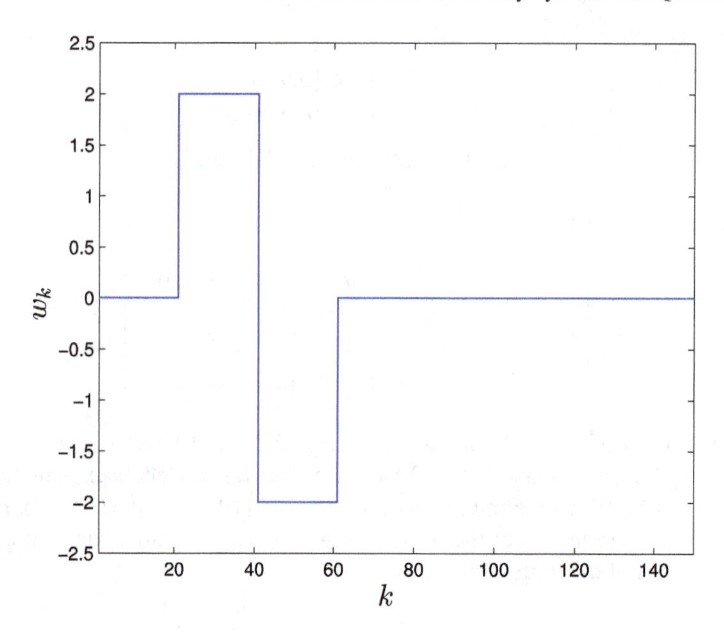

Fig. 2.3 External disturbance

Table 2.1 Different failure levels β

Case I	Case II
$\begin{bmatrix} 1 & 0 \\ 0 & 1 \end{bmatrix}$	$\begin{bmatrix} 1 & 0 \\ 0 & 0.5 \leq \beta_2 \leq 0.6 \end{bmatrix}$
Case III	Case IV
$\begin{bmatrix} 0.4 \leq \beta_1 \leq 0.5 & 0 \\ 0 & 0.5 \leq \beta_2 \leq 0.6 \end{bmatrix}$	$\begin{bmatrix} 0.1 \leq \beta_1 \leq 0.2 & 0 \\ 0 & 0.2 \leq \beta_2 \leq 0.3 \end{bmatrix}$

$$\left[\begin{array}{c|c} K_{11} & K_{21} \\ \hline K_{12} & K_{22} \\ \hline K_{13} & K_{23} \end{array}\right] = \left[\begin{array}{ccc|ccc} 0.0490 & -0.0637 & 0.0141 & 0.0283 & -0.1021 & 0.1008 \\ 0.0361 & 0.1522 & 0.0069 & -0.0718 & -0.0956 & 0.0478 \\ \hline 0.0824 & -0.1151 & 0.0339 & 0.0329 & -0.1825 & 0.1185 \\ 0.0364 & 0.1424 & 0.0059 & -0.0838 & -0.0608 & 0.0439 \\ \hline 0.0731 & -0.0987 & 0.0247 & 0.0350 & -0.1597 & 0.1173 \\ 0.0363 & 0.1478 & 0.0121 & -0.0782 & -0.0785 & 0.0465 \end{array}\right]. \quad (2.66)$$

Next, we analyze how actuator failures affect the performance of systems without the switching phenomenon. There are 4 cases of different actuator failures, shown in Table 2.1. From Table 2.2, it can been seen that when failures become worse, r^* is bigger, which implies that the system performance gets poorer.

Table 2.2 Optimal performance according to different failure levels

Case	Case I	Case II	Case III	Case IV
γ^*	2.8399	2.8417	2.8464	2.8638

2.5 Conclusion

In this chapter, we have investigated the reliable controller design problem for discrete-time T–S fuzzy systems, where randomly occurring infinite-distributed delays, the quantized control signal and actuator failures have been taken into consideration. The synchronous switching mechanisms have been introduced into the controller design and actuator failures. Then we have constructed the switched PDC controller. Sufficient conditions have been obtained to guarantee that the resulting closed-loop system is exponentially stable with the prescribed l_2–l_∞ performance index in the mean-square sense. A numerical example has been provided to validate the effectiveness of the developed approach.

References

1. Zhang, L., Ning, Z., Wang, Z.: Distributed filtering for fuzzy time-delay systems with packet dropouts and redundant channels. IEEE Trans. Syst. Man Cybern.: Syst. **46**(4), 559–572 (2016)
2. Levitin, G., Xing, L., Dai, Y.: Optimal design of hybrid redundant systems with delayed failure-driven standby mode transfer. IEEE Trans. Syst. Man Cybern.: Syst. **45**(10), 1336–1344 (2015)
3. Zhang, S., Pattipati, K.R., Hu, Z., Wen, X., Sankavaram, C.: Dynamic coupled fault diagnosis with propagation and observation delays. IEEE Trans. Syst. Man Cybern.: Syst. **43**(6), 1424–1439 (2013)
4. Zhang, S., Wang, Z., Ding, D., Shu, H.: Fuzzy filtering with randomly occurring parameter uncertainties, interval delays, and channel fadings. IEEE Trans. Cybern. **44**(3), 406–417 (2014)
5. Wei, G., Feng, G., Wang, Z.: Robust H_∞ control for discrete-time fuzzy systems with infinite-distributed delays. IEEE Trans. Fuzzy Syst. **17**(1), 224–232 (2009)
6. Wang, Z., Liu, Y., Wei, G., Liu, X.: A note on control of a class of discrete-time stochastic systems with distributed delays and nonlinear disturbances. Automatica **46**(3), 543–548 (2010)
7. Xu, S., Lam, J., Chen, T., Zou, Y.: A delay-dependent approach to robust H_∞ filtering for uncertain distributed delay systems. IEEE Trans. Signal Process. **53**(10), 3764–3772 (2005)
8. Wu, Z.-G., Shi, P., Su, H., Chu, J.: Reliable H_∞ control for discrete-time fuzzy systems with infinite-distributed delay. IEEE Trans. Fuzzy Syst. **20**(1), 22–31 (2012)
9. Wang, Z., Wei, G., Feng, G.: Reliable H_∞ control for discrete-time piecewise linear systems with infinite distributed delays. Automatica **45**(12), 2991–2994 (2009)
10. Zhang, S., Wang, Z., Ding, D., Shu, H.: H_∞ fuzzy control with randomly occurring infinite distributed delays and channel fadings. IEEE Trans. Fuzzy Syst. **22**(1), 189–200 (2014)
11. Wakaiki, M., Yamamoto, Y.: Stability analysis of sampled-data switched systems with quantization. Automatica **69**, 157–168 (2016)
12. Xing, L., Wen, C., Zhu, Y., Su, H., Liu, Z.: Output feedback control for uncertain nonlinear systems with input quantization. Automatica **65**, 191–202 (2016)
13. Yu, Z., Yan, H., Li, S., Dong, Y.: Approximation-based adaptive tracking control for switched stochastic strict-feedback nonlinear time-delay systems with sector-bounded quantization input. IEEE Trans. Syst. Man Cybern.: Syst. **48**(12), 2145–2157 (2018)

14. Yang, H., Xu, Y., Shi, P., Zhang, J.: Event-driven control for networked control systems with quantization and Markov packet losses. IEEE Trans. Cybern. **47**(8), 2235–2243 (2017)
15. Liu, Y., Wang, Z., Liang, J., Liu, X.: Synchronization and state estimation for discrete-time complex networks with distributed delays. IEEE Trans. Syst. Man Cybern. Part B: Cybern. **38**(5), 1314–1325 (2008)
16. Fu, M., Xie, L.: The sector bound approach to quantized feedback control. IEEE Trans. Autom. Control **50**(11), 1698–1711 (2005)
17. Wang, Z., Yang, Y., Ho, D.W., Liu, X.: Robust H_∞ filtering for stochastic time-delay systems with missing measurements. IEEE Trans. Signal Process. **54**(7), 2579–2587 (2006)
18. Tanaka, K., Wang, H.O.: Fuzzy Control Systems Design and Analysis: A Linear Matrix Inequality Approach. Wiley, New York (2004)

Chapter 3
Fuzzy-Model-Based Non-fragile GCC of Fuzzy MJSs

3.1 Introduction

Recent years have witnessed rising interest in the GCC problem [1, 2], whose main aim is to make a studied system stable with an adequate level of performance via devising appropriate control laws. The work in [3] has investigated the GCC issue for descriptor systems with uncertainties and robust normalization. For networked control systems, the work in [4] has focused on the output feedback GCC problem with consideration of time delays as well as stochastic packet dropouts. Through the PDC technique, the GCC issue in [5] has been studied together with quantization effects, packet dropouts and transmission delays.

Note that above mentioned works [1, 3, 4] have involved the uncertain phenomenon, which is another control problem deserving our attention. Because of inaccuracies inherent in analogue systems, finite word length in digital systems, numerical runoff during calculation and so on, running controllers or filters often experience fluctuations, which may result in the system performance degradation even with instability. Accordingly, it is of significance to devise a non-fragile filter or controller, ensuring the desired performance with powerful capabilities to tolerate admissible uncertainties. For uncertain singular state-delayed systems, the non-fragile GCC issue has been analyzed in [6] based on the PDC approach. In addition, time delays often happen in engineering applications such as mechanics, biological systems and communication systems. The work in [7] has applied the frequency discretization to discuss the delay-independent stability for linear time-invariant systems. As for switched time-varying delay systems, the work in [8] has analyzed the influence of mixed-modes on the stability analysis.

In the chapter, we focus on the non-fragile controller design for discrete-time MJSs with time-varying delays described via the T–S fuzzy approach, where the guaranteed cost index is taken into consideration. We assume that time-varying delays are arbitrary, but restricted by given upper and lower bounds. The transition probability matrix of Markov jump is also supposed to be known. We apply the idea of the PDC

S. Dong et al., *Control and Filtering of Fuzzy Systems with Switched Parameters*, Studies in Systems, Decision and Control 268,
https://doi.org/10.1007/978-3-030-35566-1_3

approach to construct the non-fragile controller. By Lyapunov–Krasovskii functions, a sufficient condition is derived for the existence of the non-fragile controller, which is expressed as LMIs. And the optimal upper bound J^* of the guaranteed cost index is also given.

3.2 Preliminary Analysis

Consider the discrete-time systems with Markov jump and time-varying delays, described by the following fuzzy rules:
Plant rule i: IF $\eta_1(k)$ is ν_{i1}, $\eta_2(k)$ is ν_{i2}, ..., and $\eta_p(k)$ is ν_{ip}, THEN

$$\begin{cases} x(k+1) = A_{\delta_k i}x(k) + A_{d\delta_k i}x(k-d(k)) + B_{\delta_k i}u(k), \\ \quad x(k) = \varphi(k),\ k \in \mathcal{S}, \end{cases} \tag{3.1}$$

where $i \in \mathcal{I} = \{1, 2, \ldots, r\}$ represents the ith fuzzy rule with total number of rules r; $\eta_j(k)$ ($j \in \chi = \{1, 2, \ldots, p\}$) is the premise variable; ν_{ij} is the fuzzy set; p and r are positive integers. These symbols $x(k) \in R^n$, $\varphi(k)$ and $u(k) \in R^m$ stand for the state vector, the initial state and the control input, respectively. $\mathcal{S} = \{-d_2, -d_2 + 1, \ldots, 0\}$ is the delay time set. The known matrices $A_{\delta_k i}$, $A_{d\delta_k i}$ and $B_{\delta_k i}$ have appropriate dimensions. The time-varying delay $d(k)$ obeys $0 < d_1 \leq d(k) \leq d_2$. The Markov jump phenomenon is described by variable $\delta_k \in \varUpsilon = \{1, 2, 3, \ldots, L\}$. The transition probability matrix is defined as $\varPi = [\pi_{st}]$ and we assume that

$$\Pr\{\delta_{k+1} = t | \delta_k = s\} = \pi_{st},\ s,\ t \in \varUpsilon, \tag{3.2}$$

where π_{st} represents the transition probability from mode s at instant k to mode t at instant $k+1$. It should be pointed out that $0 \leq \pi_{st} \leq 1$ and $\sum_{t=1}^{L} \pi_{st} = 1$.

By fuzzy inference, the overall system (3.1) is inferred as

$$x(k+1) = A_{\delta_k h}x(k) + A_{d\delta_k h}x(k-d(k)) + B_{\delta_k h}u(k), \tag{3.3}$$

where

$$\begin{aligned} A_{\delta_k h} &= \sum_{i=1}^{r} h_i(\eta(k))A_{\delta_k i},\ A_{d\delta_k h} = \sum_{i=1}^{r} h_i(\eta(k))A_{d\delta_k i}, \\ B_{\delta_k h} &= \sum_{i=1}^{r} h_i(\eta(k))B_{\delta_k i},\ h_i(\eta(k)) = \frac{\prod_{j=1}^{p} \nu_{ij}(\eta_j(k))}{\sum_{i=1}^{r}\prod_{j=1}^{p} \nu_{ij}(\eta_j(k))}, \\ \eta(k) &= [\eta_1(k), \eta_2(k), \ldots, \eta_p(k)]. \end{aligned}$$

The variable $\nu_{ij}(\eta_j(k))$ denotes the grade of the membership of $\eta_j(k)$ in ν_{ij}, and satisfies $\prod_{j=1}^{p}\nu_{ij}(\eta_j(k)) \geq 0$. $h_i(\eta(k))$ is the normalized fuzzy weighting function with

$$h_i(\eta(k)) \geq 0,\ i \in \mathcal{I}, \tag{3.4}$$

and

$$\sum_{i=1}^{r} h_i(\eta(k)) = 1. \tag{3.5}$$

In order to analyze conveniently, $h_i(\eta(k))$ is abbreviated as h_i later.

Remark 3.1 Generally, a T–S fuzzy logic system [9] is made up of four basis parts, namely, the fuzzier, the fuzzy rule base, the inference engine and the defuzzifier. There are two main approaches to achieve the T–S fuzzy logic system: (1) the derivation for the given nonlinear systems, which uses the ideas of the sector nonlinearity, the local approximation or even both; (2) the identification technique containing the parameter identification and the structure identification. It is worth noting that membership functions have an important effect on the system performance, and the work in [10] has investigated them deeply.

Remark 3.2 The nonlinear MJSs are described via the T–S fuzzy method that is an overall approximator. And every local linear system, called as a subsystem, is derived as $A_{\delta_k i}x(k) + A_{d\delta_k i}x(k-d(k)) + B_{\delta_k i}u(k)$. Then through an average weighted sum of local linear subsystems, we have system (3.3). Here, to avoid complex calculation of the defuzzification process in designing the fuzzy controller, we suppose that premise variable $\eta_j(k)$ is independent of the control input $u(k)$.

Due to uncertainties of the controller implementation, the following non-fragile fuzzy controller is devised via using the PDC approach, as follows:
Controller rule i: IF $\eta_1(k)$ is ν_{i1}, $\eta_2(k)$ is ν_{i2}, ..., and $\eta_p(k)$ is ν_{ip}, THEN

$$u(k) = -(K_i + \Delta K_i)x(k), \tag{3.6}$$

where $K_i \in R^{m\times n}$ is the state feedback controller matrix of the ith subfuzzy, and it is to be determined. ΔK_i denotes the drift of the controller and $\Delta K_i = E_i F_i(k) H_i$, where given matrices E_i and H_i have appropriate dimensions. The uncertain matrix $F_i(k)$ is time-varying with $F_i^T(k)F_i(k) < I$.

From (3.6), it follows that

$$u(k) = -\bar{K}_h x(k), \tag{3.7}$$

where

$$\bar{K}_h = K_h + \Delta K_h = \sum_{i=1}^{r} h_i \bar{K}_i,$$
$$K_h = \sum_{i=1}^{r} h_i K_i, \ \Delta K_h = \sum_{i=1}^{r} h_i E_i F_i(k) H_i,$$
$$\bar{K}_i = K_i + E_i F_i(k) H_i.$$

Considering (3.3) and (3.7) with $\delta_k = s$, the closed-loop fuzzy system is obtained:

$$x(k+1) = \bar{A}_{sh} x(k) + A_{dsh} x(k - d(k)), \quad (3.8)$$

where

$$\bar{A}_{sh} = A_{sh} - B_{sh} \bar{K}_h = \sum_{i=1}^{r} \sum_{j=1}^{r} h_i h_j \bar{A}_{sij},$$
$$A_{sh} = \sum_{i=1}^{r} h_i A_{si}, \ B_{sh} \bar{K}_h = \sum_{i=1}^{r} \sum_{j=1}^{r} h_i h_j B_{si} \bar{K}_j,$$
$$A_{dsh} = \sum_{i=1}^{r} h_i A_{dsi}, \ \bar{A}_{sij} = A_{si} - B_{si} \bar{K}_j.$$

Remark 3.3 Generally speaking, there are two types of Markov chains, namely, homogeneous and nonhomogeneous chains. Here, we apply the homogeneous Markov process to describe hybrid systems. Since it is quite difficult to obtain sufficient and efficient information about the operational system modes, we focus on the mode-independent controller design, shown in (3.6).

The guaranteed cost index is introduced for constructing a guaranteed cost controller (3.6) as below:

$$J = \sum_{k=1}^{\infty} E\left\{ x^T(k) Q_1 x(k) + u^T(k) Q_2 u(k) \right\}, \quad (3.9)$$

where Q_1 and Q_2 are known positive-definite symmetric matrices.

We present the following assumption and definition, which are necessary for investigation later.

Assumption 3.1 *The initial state $x(d)$ belongs to a set:*

$$N = \{x(d) \in R^n : \ x(d) = M\bar{x}(d), \ \bar{x}^T(d)\bar{x}(d) \le 1\},$$

where M is a given matrix (or a matrix to be determined) and $d \in \mathcal{D} = \{0, -1, -2, -3, \ldots\}$.

Definition 3.1 ([3]) When there exists a positive scalar J^* and a fuzzy control law (3.6) guaranteeing that system (3.8) maintains asymptotical stability with $J \leq J^*$ for the cost function (3.9), J^* is the upper bound of the GCC index and $u(k)$ is the non-fragile GCC law for system (3.3).

This chapter's goal is to construct the controller (3.6) that satisfies two requirements in the following:
(1) The closed-loop fuzzy system (3.8) is stochastically stable;
(2) As for system (3.8) and controller (3.6), the cost index (3.9) meets $J \leq J^*$, where J^* is the optimal upper bound of the non-fragile GCC.

3.3 Main Results

In this section, we start with developing a sufficient condition that makes the system (3.8) stochastically stable with an upper bound of the GCC index on the supposition that the controller gain K_i in (3.6) is given. Then we devote ourselves to investigating the solution to K_i and J^*.

Theorem 3.4 *If there exist matrices $P_{si} > 0$, $R_{zi} > 0$ for any $s,\ t \in \Upsilon, a,\ b,\ c,\ f,\ i,\ j\ \in \mathcal{I}$ and $z \in \{1, 2, 3\}$ satisfying*

$$\Xi_{sabcfii} < 0, \tag{3.10}$$

$$\Xi_{sabcfij} + \Xi_{sabcfji} < 0, i < j, \tag{3.11}$$

system (3.8) is stochastically stable with the guaranteed cost index J satisfying

$$\begin{aligned} J &< x^T(0)P_s(0)x(0) + \sum_{\tau=-d_1}^{-1} x^T(\tau)R_1(\tau)x(\tau) \\ &+ \sum_{\tau=-d_2}^{-1-d_1} x^T(\tau)R_2(\tau)x(\tau) + \sum_{\gamma=-d_2+1}^{-d_1+1} \sum_{\tau=\gamma-1}^{-1} x^T(\tau)R_3(\tau)x(\tau), \end{aligned} \tag{3.12}$$

where

$$\Xi_{sabcfij} = \begin{bmatrix} \Xi_{11} & * \\ \Xi_{21} & \Xi_{22} \end{bmatrix},\ \Xi_{11} = diag\{\Xi_{11}^{11}, -R_{1a} + R_{2a}, -R_{3b}, -R_{2c}\},$$
$$\Xi_{21} = \begin{bmatrix} \phi_{1sij}^T \bar{P}_{sf} \ \phi^T \ \phi_{2j}^T \end{bmatrix}^T,\ \Xi_{22} = diag\{-\bar{P}_{sf}, -Q_1^{-1}, -Q_2^{-1}\},$$
$$\Xi_{11}^{11} = -P_{si} + R_{1i} + (d_{12} + 1)R_{3i},\ \phi = \begin{bmatrix} I_n \ 0\ 0\ 0 \end{bmatrix},\ \phi_{1sij} = \begin{bmatrix} \bar{A}_{sij}\ 0\ A_{dsi}\ 0 \end{bmatrix},$$
$$\phi_{2j} = \begin{bmatrix} \bar{K}_j\ 0\ 0\ 0 \end{bmatrix},\ \bar{P}_{sf} = \sum_{t=1}^{L} \pi_{st} P_{tf},\ d_{12} = d_2 - d_1.$$

Proof First, define

$$
\begin{aligned}
&P_s(k) = \sum_{i=1}^{r} h_i P_{si},\ P_t(k+1) = \sum_{f=1}^{r} h_f P_{tf},\\
&R_1(k) = \sum_{i=1}^{r} h_i R_{1i},\ R_1(k-d_1) = \sum_{a=1}^{r} h_a R_{1a},\\
&R_2(k) = \sum_{i=1}^{r} h_i R_{2i},\ R_2(k-d_2) = \sum_{c=1}^{r} h_c R_{2c},\\
&R_3(k) = \sum_{i=1}^{r} h_i R_{3i},\ R_3(k-d(k)) = \sum_{b=1}^{r} h_b R_{3b},\\
&\bar{P}_s(k+1) = \sum_{t=1}^{L} \pi_{st} P_t(k+1) = \sum_{f=1}^{r}\sum_{t=1}^{L} h_f \pi_{st} P_{tf} = \sum_{f=1}^{r} h_f \bar{P}_{sf}.
\end{aligned}
\tag{3.13}
$$

From (3.10) together with (3.7) and (3.8), we have

$$
\begin{aligned}
\Xi(k) =& \sum_{a=1}^{r}\sum_{b=1}^{r}\sum_{c=1}^{r}\sum_{f=1}^{r}\sum_{i=1}^{r}\sum_{j=1}^{r} h_a h_b h_c h_f h_i h_j \Xi_{sabcfij}\\
=& \sum_{a=1}^{r}\sum_{b=1}^{r}\sum_{c=1}^{r}\sum_{f=1}^{r}\sum_{j=1}^{r} h_a h_b h_c h_f \left(\sum_{i=1}^{r} h_i^2 \Xi_{sabcfii}\right.\\
&\left. + \sum_{i=1}^{r-1}\sum_{j=i+1}^{r} h_i h_j (\Xi_{sabcfij} + \Xi_{sabcfji})\right) < 0,
\end{aligned}
\tag{3.14}
$$

where

$$
\begin{aligned}
&\Xi(k) = \begin{bmatrix} \Xi'_{11} & * \\ \Xi'_{21} & \Xi'_{22} \end{bmatrix},\\
&\Xi'_{11} = \mathrm{diag}\{\Xi_{11}^{'11}, \Xi_{11}^{'22}, -R_3(d-d(k)), -R_2(k-d_2)\},\\
&\Xi'_{21} = \left[\phi_{1s}^T(k)\bar{P}_s(k+1)\ \phi^T\ \phi_2^T(k)\right]^T,\\
&\Xi'_{22} = \mathrm{diag}\{-\bar{P}_s(k+1), -Q_1^{-1}, -Q_2^{-1}\},\\
&\Xi_{11}^{'11} = -P_s(k) + R_1(k) + (d_{12}+1)R_3(k),\\
&\Xi_{11}^{'22} = -R_1(k-d_1) + R_2(k-d_1),\\
&\phi_{1s}(k) = \left[\bar{A}_{sh}\ 0\ A_{dsh}\ 0\right],\ \phi_2(k) = \left[\bar{K}_h\ 0\ 0\ 0\right].
\end{aligned}
$$

According to Schur Complement, it follows from $\Xi(k) < 0$ that

$$
\Xi'_{11} - \Xi_{21}^{'T}\Xi_{22}^{'-1}\Xi'_{21} < 0,
\tag{3.15}
$$

which equals to

$$\Xi'(k) = \phi_{1s}^T(k)\bar{P}_s(k+1)\phi_{1s}(k) + \Xi''_{11} < 0, \tag{3.16}$$

where

$$\Xi''_{11} = \text{diag}\{\Xi''^{11}_{11}, \Xi'^{22}_{11}, -R_3(d-d(k)), -R_2(k-d_2)\},$$
$$\Xi''^{11}_{11} = -P_s(k) + R_1(k) + (d_{12}+1)R_3(k) + Q_1 + \bar{K}_h^T Q_2 \bar{K}_h.$$

The following Lyapunov–Krasovskii functions are adopted:

$$V(k) = \sum_{i=1}^{4} V_i(k), \tag{3.17}$$

where

$$V_1(k) = x^T(k)P_s(k)x(k), \;\; V_2(k) = \sum_{\tau=k-d_1}^{k-1} x^T(\tau)R_1(\tau)x(\tau),$$
$$V_3(k) = \sum_{\tau=k-d_2}^{k-1-d_1} x^T(\tau)R_2(\tau)x(\tau), \;\; V_4(k) = \sum_{\gamma=-d_2+1}^{-d_1+1} \sum_{\tau=k+\gamma-1}^{k-1} x^T(\tau)R_3(\tau)x(\tau).$$

Compute the difference of every term in $V(k)$ with system (3.8), and then take the expectation, as follows:

$$\begin{aligned}
E\{\Delta V_1(k)\} &= E\{x^T(k+1)P_t(k+1)x_(k+1)\} - E\{x^T(k)P_s(k)x(k)\} \\
&= \xi^T(k)\phi_{1s}^T(k)\bar{P}_s(k+1)\phi_{1s}(k)\xi(k) - x^T(k)P_s(k)x(k), \\
E\{\Delta V_2(k)\} &= x^T(k)R_1(k)x(k) - x^T(k-d_1)R_1(k-d_1)x(k-d_1), \\
E\{\Delta V_3(k)\} &= x^T(k-d_1)R_2(k-d_1)x(k-d_1) \\
&\quad - x^T(k-d_2)R_2(k-d_2)x(k-d_2),
\end{aligned} \tag{3.18}$$

$$\begin{aligned}
E\{\Delta V_4(k)\} &= (d_{12}+1)x^T(k)R_3(k)x(k) - \sum_{\gamma=k-d_2}^{k-d_1} x^T(\gamma)R_3(\gamma)x(\gamma) \\
&\le (d_{12}+1)x^T(k)R_3(k)x(k) \\
&\quad - x^T(k-d(k))R_3(k-d(k))x(k-d(k)),
\end{aligned} \tag{3.19}$$

where

$$\xi(k) = \left[x^T(k) \; x^T(k-d_1) \; x^T(k-d(k)) \; x^T(k-d_2)\right]^T.$$

Accordingly,

$$E\{\Delta V(k) + x^T(k)(Q_1 + \bar{K}_h^T Q_2 \bar{K}_h)x(k)\} \le \xi_k^T \Xi'(k)\xi_k. \tag{3.20}$$

Owing to $\Xi'(k) < 0$, $Q_1 > 0$, and $Q_2 > 0$, it can conclude that

$$E\{\Delta V(k)\} < 0, \tag{3.21}$$

and

$$E\left\{x^T(k)(Q_1 + \bar{K}_h^T Q_2 \bar{K}_h)x(k)\right\} < -E\{\Delta V(k)\}. \tag{3.22}$$

Inequality (3.21) implies that the closed-loop fuzzy system (3.8) is stochastically stable.

Adding up both sides of (3.22) from $k = 0$ to ∞, we obtain that

$$E\left\{\sum_{k=0}^{\infty} x^T(k)(Q_1 + \bar{K}_h^T Q_2 \bar{K}_h)x(k)\right\} < E\{V(0)\}. \tag{3.23}$$

Accordingly, the non-fragile GCC function (3.9) satisfies (3.12). The proof is complete.

Remark 3.5 The number of terms in Lyapunov–Krasovskii function is fewer than that in [11, 12], which reduces the number of unknown matrices and brings less computation. Besides, $P_s(k)$ in Lyapunov–Krasovskii functional (3.17) is not only fuzzy-basis-dependent, but also mode-dependent. Moreover, $R_1(k)$, $R_2(k)$, and $R_3(k)$ are fuzzy-basis-dependent as well. Comparing with common Lyapunov functions, it leads to less conservatism.

Now, in the light of Theorem 3.4, the next step is to investigate the solution to the controller gain.

Theorem 3.6 *If there exist matrices $\breve{P}_{si} > 0$, $\breve{R}_{zi} > 0$, Z, $\breve{K}_i$, scalars $\varepsilon_z > 0$ for any $s, t \in \Upsilon$, $a, b, c, f, i, j \in \mathcal{I}$ and $z \in \{1, 2, 3\}$ satisfying*

$$\begin{bmatrix} \Psi^1_{sabcfii} & * & * \\ \Psi^2_i & -\varepsilon_1 I & * \\ \varepsilon_1 \Psi^3_{sii} & 0 & -\varepsilon_1 I \end{bmatrix} < 0, \tag{3.24}$$

$$\begin{bmatrix} \Psi^1_{sabcfij} + \Psi^1_{sabcfji} & * & * & * & * \\ \Psi^2_j & -\varepsilon_2 I & * & * & * \\ \varepsilon_2 \Psi^3_{sij} & 0 & -\varepsilon_2 I & * & * \\ \Psi^2_i & 0 & 0 & -\varepsilon_3 I & * \\ \varepsilon_3 \Psi^3_{sji} & 0 & 0 & 0 & -\varepsilon_3 I \end{bmatrix} < 0,\ i < j, \tag{3.25}$$

system (3.8) is stochastically stable with an upper bound of the GCC index in (3.12). The controller parameter can be derived as

$$K_j = \breve{K}_j Z^{-1}, \tag{3.26}$$

where

$$\Psi^1_{sabcfij} = \begin{bmatrix} \Psi_{11} & * \\ \Psi_{21} & \Psi_{22} \end{bmatrix}, \ \Psi^2_j = \begin{bmatrix} H_j Z \ 0 \ 0 \ 0 \ 0 \ 0 \ 0 \end{bmatrix},$$
$$\Psi^3_{sij} = E^T_j \begin{bmatrix} 0 \ 0 \ 0 \ 0 \ -B^T_{si} \ 0 \ I_m \end{bmatrix},$$
$$\Psi_{11} = diag\{\Psi^{11}_{11}, -\breve{R}_{1a} + \breve{R}_{2a}, -\breve{R}_{3b}, -\breve{R}_{2c}\},$$
$$\Psi_{21} = \begin{bmatrix} \phi'^T_{1sij} \ \phi'^T \ \phi'^T_{2j} \end{bmatrix}^T, \ \Psi_{22} = diag\{\breve{\bar{P}}_{sf} - Z - Z^T, -Q^{-1}_1, -Q^{-1}_2\},$$
$$\Psi^{11}_{11} = -\breve{P}_{si} + \breve{R}_{1i} + (d_{12} + 1)\breve{R}_{3i}, \ \phi' = \begin{bmatrix} Z \ 0 \ 0 \ 0 \end{bmatrix},$$
$$\phi'_{1sij} = \begin{bmatrix} A_{si}Z - B_{si}\breve{K}_j \ 0 \ A_{dsi}Z \ 0 \end{bmatrix}, \ \phi'_{2j} = \begin{bmatrix} \breve{K}_j \ 0 \ 0 \ 0 \end{bmatrix},$$
$$\breve{\bar{P}}_{sf} = \sum_{t=1}^{L} \pi_{st}\breve{P}_{tf}, \ d_{12} = d_2 - d_1.$$

Proof By Schur Complement, it follows from (3.24) that

$$\Psi^1_{sabcfii} + \varepsilon^{-1}_1 \Psi^{2T}_i \Psi^2_i + \varepsilon_1 \Psi^{3T}_{sii} \Psi^3_{sii} < 0. \tag{3.27}$$

Owing to $F^T_i(k)F_i(k) < I$, the following inequality is obtained:

$$\Psi^1_{sabcfii} + \Psi^{3T}_{sii} F_i(k)\Psi^2_i + \Psi^{2T}_i F^T_i(k)\Psi^3_{sii} < 0. \tag{3.28}$$

The above inequality can be rewritten as

$$\Psi^{1'}_{sabcfii} < 0, \tag{3.29}$$

where

$$\Psi^{1'}_{sabcfii} = \begin{bmatrix} \Psi_{11} & * \\ \Psi'_{21} & \Psi_{22} \end{bmatrix}, \ \Psi'_{21} = \begin{bmatrix} \phi^T_{1sii} \ \phi^T \ \phi^T_{2i} \end{bmatrix}^T Z, \ \breve{K}_i = K_i Z.$$

Due to $\bar{P}_{sf} = \sum_{t=1}^{L} \pi_{st} P_{tf} > 0$, we have

$$(\bar{P}^{-1}_{sf} - Z^T)\bar{P}_{sf}(\bar{P}^{-1}_{sf} - Z) > 0. \tag{3.30}$$

Hence,

$$-\bar{P}^{-1}_{sf} < \breve{\bar{P}}_{sf} - Z - Z^T, \tag{3.31}$$

where $\breve{\bar{P}}_{sf} = Z^T \bar{P}_{sf} Z = \sum_{t=1}^{L} \pi_{st}\breve{P}_{tf}$, $\breve{P}_{tf} = Z^T P_{tf} Z$.

The similar way is employed to deal with matrices P_{si} and R_{zi}. Then, we obtain that

$$\begin{cases} -P^{-1}_{si} < \breve{P}_{si} - Z - Z^T, \\ -R^{-1}_{zi} < \breve{R}_{zi} - Z - Z^T, \end{cases} \tag{3.32}$$

where

$$\breve{P}_{si} = Z^T P_{si} Z,\ \breve{R}_{zi} = Z^T R_{zi} Z,\ z \in \{1, 2, 3\}.$$

According to (3.29) and (3.31), we can infer that

$$\Psi^{1''}_{sabcfii} < 0, \tag{3.33}$$

where

$$\Psi^{1''}_{sabcfii} = \begin{bmatrix} \Psi_{11} & * \\ \Psi'_{21} & \Psi'_{22} \end{bmatrix},\ \Psi'_{22} = \text{diag}\{-\bar{P}^{-1}_{sf}, -Q_1^{-1}, -Q_2^{-1}\}.$$

By pre-multiplying $\text{diag}\{Z^T, Z^T, Z^T, Z^T, \bar{P}^{-1}_{sf}, I, I\}$ and post-multiplying $\text{diag}\{Z, Z, Z, Z, \bar{P}^{-1}_{sf}, I, I\}$ to (3.10) respectively, (3.33) is achieved. Hence, we easily conclude that (3.10) is guaranteed by (3.24). In the similar way, it is clearly observed that we can obtain (3.11) from (3.25). If (3.24) and (3.25) are feasible, the controller gain can be derived as $K_j = \breve{K}_j Z^{-1}$. The proof is complete.

Theorem 3.6 offers a solution to the fuzzy controller and the upper bound of non-fragile guaranteed cost index. We can acutely observe that the upper bound of J is related to initial states. Therefore, we desire to find the optimal upper bound of J, which is independent of initial states.

Considering Assumption 3.1, the initial state $x(d)$ satisfies

$$x(d) = M\bar{x}(d),\ \bar{x}^T(d)\bar{x}(d) \leq 1, d \in \mathcal{D}.$$

Hence, the following inequalities are obtained:

$$\begin{aligned} & x^T(0)P_s(0)x(0) \leq \lambda_{max}(M^T P_{si} M), \\ & \sum_{\tau=-d_1}^{-1} x^T(\tau)R_1(\tau)x(\tau) \leq d_1\lambda_{max}(M^T R_{1i} M), \\ & \sum_{\tau=-d_2}^{-1-d_1} x^T(\tau)R_2(\tau)x(\tau) \leq d_{12}\lambda_{max}(M^T R_{2i} M), \\ & \sum_{\gamma=-d_2+1}^{-d_1+1} \sum_{\tau=\gamma-1}^{-1} x^T(\tau)R_3(\tau)x(\tau) \leq d_3\lambda_{max}(M^T R_{3i} M), \end{aligned} \tag{3.34}$$

where

$$d_3 = \frac{(d_1 + d_2)(d_{12} + 1)}{2}.$$

Proof Here the proof of the second inequality in (3.34) is given as follows.

Based on Assumption 3.1 and $R_{1i} > 0$, it follows that

$$\begin{aligned} x^T(\tau)R_1(\tau)x(\tau) &= \bar{x}^T(\tau)M^T R_1(\tau)M\bar{x}(\tau) \\ &= \sum_{i=1}^{r} h_i(\tau)\bar{x}^T M^T R_{1i} M\bar{x}(\tau) \\ &\leq \sum_{i=1}^{r} h_i(\tau)\lambda_{max}(M^T R_{1i} M)\bar{x}^T(\tau)\bar{x}(\tau) \\ &\leq \sum_{i=1}^{r} h_i(\tau)\lambda_{max}(M^T R_{1i} M). \end{aligned} \tag{3.35}$$

Due to $\sum_{i=1}^{r} h_i(\eta(k)) = 1$ in (3.5), we have

$$x^T(\tau)R_1(\tau)x(\tau) \leq \lambda_{max}(M^T R_{1i} M). \tag{3.36}$$

Accordingly,

$$\sum_{\tau=-d_1}^{-1} x^T(\tau)R_1(\tau)x(\tau) \leq d_1\lambda_{max}(M^T R_{1i} M). \tag{3.37}$$

The other inequalities in (3.34) follow the similar process above, which are omitted owing to space limitation.

We introduce scalars $\alpha_w > 0$ ($w \in \{0, 1, 2, 3\}$) such that

$$\begin{cases} \lambda_{max}(M^T P_{si} M) < \alpha_0, \\ \lambda_{max}(M^T R_{1i} M) < \alpha_1, \\ \lambda_{max}(M^T R_{2i} M) < \alpha_2, \\ \lambda_{max}(M^T R_{3i} M) < \alpha_3. \end{cases} \tag{3.38}$$

By Schur Complement, we find that

$$\begin{cases} \begin{bmatrix} -\alpha_0 I & * \\ -M & -P_{si}^{-1} \end{bmatrix} < 0, \\ \begin{bmatrix} -\alpha_1 I & * \\ -M & -R_{1i}^{-1} \end{bmatrix} < 0, \\ \begin{bmatrix} -\alpha_2 I & * \\ -M & -R_{2i}^{-1} \end{bmatrix} < 0, \\ \begin{bmatrix} -\alpha_3 I & * \\ -M & -R_{3i}^{-1} \end{bmatrix} < 0. \end{cases} \tag{3.39}$$

Taking (3.32) into consideration, (3.39) is guaranteed by the following inequalities:

$$\begin{cases} \begin{bmatrix} -\alpha_0 I & * \\ -M & \breve{P}_{si} - Z - Z^T \end{bmatrix} < 0, \\ \begin{bmatrix} -\alpha_1 I & * \\ -M & \breve{R}_{1i} - Z - Z^T \end{bmatrix} < 0, \\ \begin{bmatrix} -\alpha_2 I & * \\ -M & \breve{R}_{2i} - Z - Z^T \end{bmatrix} < 0, \\ \begin{bmatrix} -\alpha_3 I & * \\ -M & \breve{R}_{3i} - Z - Z^T \end{bmatrix} < 0. \end{cases} \tag{3.40}$$

Define

$$J^* = \alpha_0 + d_1\alpha_1 + d_{12}\alpha_2 + d_3\alpha_3. \tag{3.41}$$

Accordingly,

$$J \leq J^*. \tag{3.42}$$

It is easily observed that we can achieve the optimal upper bound of J via minimizing J^*, which is the following theorem.

Theorem 3.7 *If there exist matrices $\breve{P}_{si} > 0$, $\breve{R}_{zi} > 0$, Z, $\breve{K}_i$, scalars $\alpha_0 > 0$, $\alpha_z > 0$, $\varepsilon_z > 0$ for any $s,\ t \in \Upsilon,\ a,\ b,\ c,\ f,\ i,\ j\ \in \mathcal{I}$ and $z \in \{1, 2, 3\}$ such that the convex optimization problem in the following can be solved:*

$$\begin{aligned} &\mathit{min} && J^* = \alpha_0 + d_1\alpha_1 + d_{12}\alpha_2 + d_3\alpha_3, \\ &\mathit{subject\ to} && (3.24),\ (3.25)\ \mathit{and}\ (3.40), \end{aligned}$$

system (3.8) is stochastically stable with the control law (3.7). The controller parameter can be derived as

$$K_j = \breve{K}_j Z^{-1}. \tag{3.43}$$

The guaranteed cost index J has the optimal upper bound J^*.

Remark 3.8 In Theorem 3.7, the GCC issue is transformed into a convex optimization problem. By applying LMI Toolbox in Matlab to solve (3.24), (3.25) and (3.40), the non-fragile controller gain (3.7) with the optimal upper bound of the cost performance (3.41) can be derived.

3.4 Illustrative Example

To validate the effectiveness and correctness of our developed approach, a single-link robot arm system [11, 12] is provided, as follows:

$$\begin{cases} \dot{x}_1(t) = f x_2(t) + (1-f) x_2(t-d(t)), \\ \dot{x}_2(t) = -\dfrac{glM}{J}\sin(x_1(t)) - \dfrac{fR}{J}x_2(t) \\ \qquad - \dfrac{(1-f)R}{J}x_2(t-d(t)) + \dfrac{1}{J}u(t), \end{cases}$$

where $x_1(t)$ and $x_2(t)$ represent the angel and angular velocity of the robot arm, respectively. $f,\ g,\ M,\ J,\ l,$ and R are the retarded coefficient, the gravity acceleration, the payload mass, the inertia moment, the robot arm length and the viscous friction coefficient, respectively. It is assumed that $R = 2,\ l = 0.5,$ and $g = 9.81$. Moreover, as time goes by, M and J vary among three modes: $M_1 = J_1 = 1,\ M_2 = J_2 = 5,$ and $M_3 = J_3 = 10$, which lead to jumps of the considered system. The transition probability matrix of Markov jump is further supposed to be

$$\Pi = \begin{bmatrix} 0.6 & 0.2 & 0.2 \\ 0.1 & 0.8 & 0.1 \\ 0.2 & 0.2 & 0.6 \end{bmatrix}.$$

After discretization with the fuzzy inference to the original system, the discrete-time fuzzy MJSs is attained, as follows:
Plant rule 1: IF $x_1(k)$ is about 0 rad, THEN

$$x(k+1) = A_{\delta_k 1}x(k) + A_{d\delta_k 1}x(k-d(k)) + B_{\delta_k 1}u(k),$$

Plant rule 2: IF $x_1(k)$ is about π rad or $-\pi$ rad, THEN

$$x(k+1) = A_{\delta_k 2}x(k) + A_{d\delta_k 2}x(k-d(k)) + B_{\delta_k 2}u(k),$$

where

$$A_{\delta_k 1} = \begin{bmatrix} 1 & Tf \\ -\frac{TglM_{\delta_k}}{J_{\delta_k}} & 1-\frac{TfR}{J_{\delta_k}} \end{bmatrix},\ A_{\delta_k 2} = \begin{bmatrix} 1 & Tf \\ -\frac{\beta TglM_{\delta_k}}{J_{\delta_k}} & 1-\frac{TfR}{J_{\delta_k}} \end{bmatrix},$$

$$A_{d\delta_k 1} = A_{d\delta_k 2} = \begin{bmatrix} 0 & T(1-f) \\ 0 & -\frac{T(1-f)R}{J_{\delta_k}} \end{bmatrix},\ B_{\delta_k 1} = B_{\delta_k 2} = \begin{bmatrix} 0 \\ \frac{T}{J_{\delta_k}} \end{bmatrix},\ \delta_k = \{1, 2, 3\},$$

$x_1(k) \in (-\pi,\ \pi)$, $T = 0.1$, and $\beta = 10^{-2}/\pi$.

The normalized fuzzy weighting functions are defined as

$$h_1(x_1(k)) = \begin{cases} \dfrac{\sin(x_1(k)) - \beta x_1(k)}{(1-\beta)x_1(k)}, & x_1(k) \neq 0, \\ 1, & x_1(k) = 0, \end{cases}$$

$$h_2(x_1(k)) = 1 - h_2(x_1(k)).$$

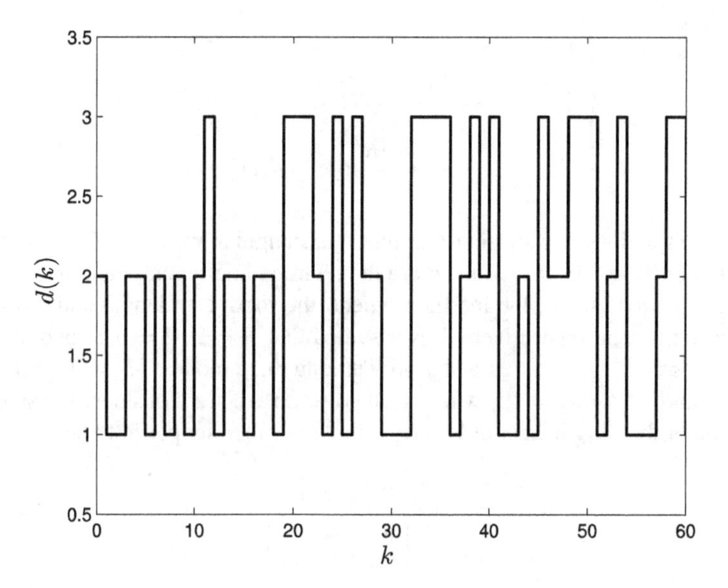

Fig. 3.1 Time-varying delays

The matrices Q_1 and Q_2 of the non-fragile guaranteed cost index (3.9) are given as

$$Q_1 = \begin{bmatrix} 0.5 & 0 \\ 0 & 0.5 \end{bmatrix}, \quad Q_2 = 1.$$

We suppose that $d(k) \in \{1, 2, 3\}$ is plotted in Fig. 3.1. Initial states are assumed to be $x(0) = \begin{bmatrix} 0.5\pi & -0.5\pi \end{bmatrix}^T$ and $x(\sigma) = \begin{bmatrix} 0 & 0 \end{bmatrix}^T$ $(\sigma \in \{-1, -2, -3\})$. Hence, based on Assumption 3.1, we choose M as

$$M = \begin{bmatrix} 2.5 & 0 \\ 0 & 2.5 \end{bmatrix}.$$

The drift matrices of the devised controller (3.7) are

$$\begin{aligned} &E_1 = 0.02, \ E_2 = -0.02, \\ &H_1 = \begin{bmatrix} 0.02 & 0 \end{bmatrix}, \ H_2 = \begin{bmatrix} 0 & -0.01 \end{bmatrix}, \\ &F_1(k) = 0.5\sin(k), \ F_2(k) = -0.5\cos(k). \end{aligned}$$

Then we apply LMI Toolbox to solve (3.24), (3.25) and (3.40), and obtain Z and $\breve{K}_i (i = 1, \ 2)$:

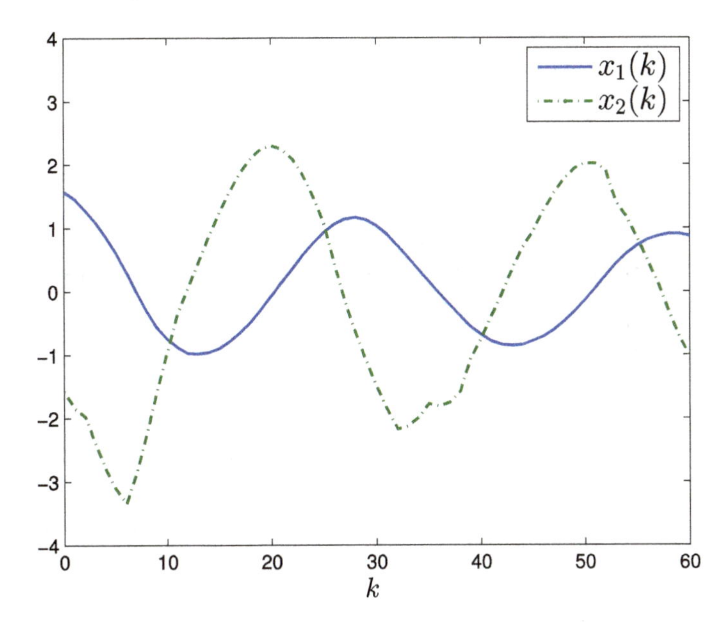

Fig. 3.2 State responses of the system (3.3)

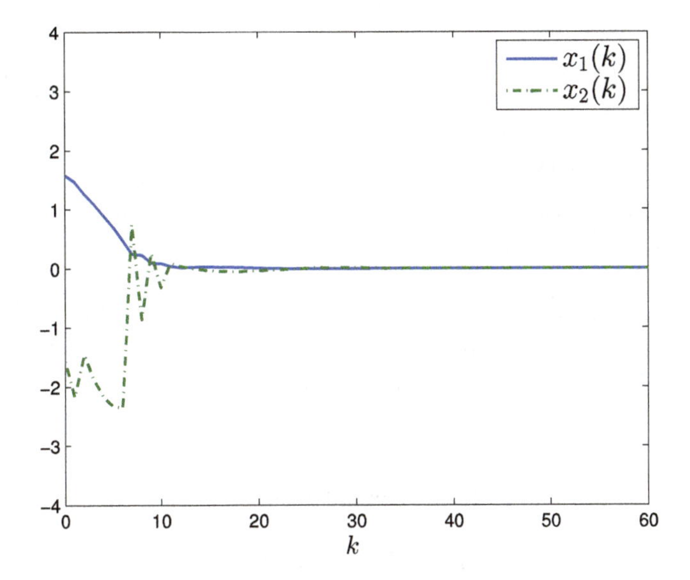

Fig. 3.3 State responses of the system (3.8)

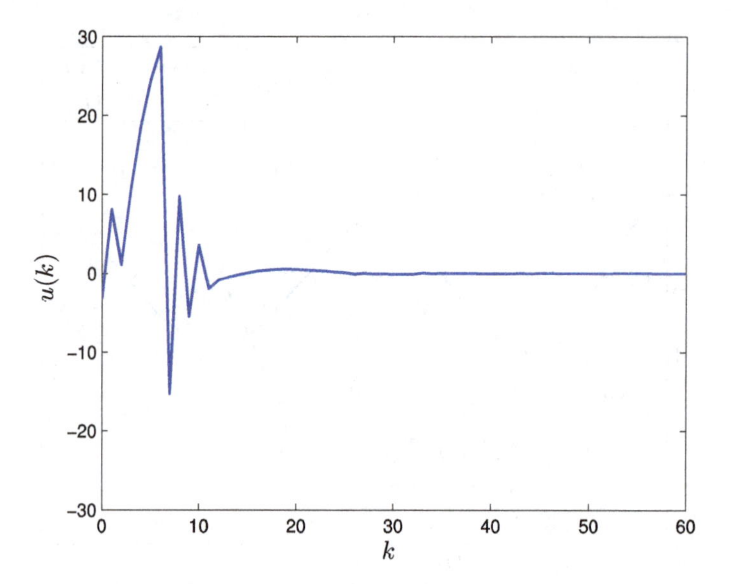

Fig. 3.4 Control input for the system (3.3)

Table 3.1 Optimal upper cost value J^* for different upper bound d_2

d_2	2	3	4	5	6
J^* ($\times 10^5$)	0.8486	1.7576	3.3607	6.1793	11.3186

Table 3.2 Optimal upper cost value J^* for different lower bound d_1

d_1	3	4	5	6	7
J^* ($\times 10^5$)	16.6938	11.1286	7.4461	4.8379	2.9265

$$Z = 10^{-3} \times \begin{bmatrix} 0.0388 & -0.0498 \\ -0.0554 & 0.2784 \end{bmatrix},$$
$$\breve{K}_1 = 10^{-4} \times \begin{bmatrix} -2.5 & 35.0 \end{bmatrix}, \ \breve{K}_2 = 10^{-4} \times \begin{bmatrix} -0.4 & 31.4 \end{bmatrix}.$$

Based on $K_i = \breve{K}_i Z^{-1}$, non-fragile controller gains are achieved as follows:

$$K_1 = \begin{bmatrix} 15.6044 & 15.3637 \end{bmatrix}, \ K_2 = \begin{bmatrix} 20.1963 & 14.8724 \end{bmatrix}.$$

In simulation, the retarded coefficient is supposed to be $f = 0.7$. From Fig. 3.2, we can easily observe that system (3.3) is unstable in [0, 60]. Figure 3.3, under the non-fragile control input in Fig. 3.4, presents that states $x_1(k)$ and $x_2(k)$ tend to zero over time. By calculating, the minimal upper bound of J is obtained: $J^* = 1.2864 \times 10^6$. These results confirm the applicability of our developed technique.

Now, we are interested in analyzing the relationship between optimal upper guaranteed cost J^* and time-varying delay $d(k)$. At this time, we assume that the retarded coefficient is $f = 0.8$. Firstly, the lower bound d_1 is assumed to be fixed at 1 while the upper bound d_2 varies. Via LMI Toolbox, the corresponding J^* is obtained. More detailed outcomes are presented in Table 3.1, where we can clearly note that when d_1 is fixed, J^* rises with the increase of d_2. When d_2 is fixed at 8, we change d_1. The optimal upper guaranteed cost value J^* is obtained in Table 3.2. We can clearly find out that the bigger d_1 is, the smaller J^* is. In conclusion, the time-varying delays have a significant impact on the system performance, deserving our further investigation.

3.5 Conclusion

We have developed one approach for dealing with the non-fragile GCC issue for fuzzy MJSs with time-varying delays in this chapter. The developed technique can maintain that the closed-loop system is stochastically stable. Furthermore, the optimal upper bound of the cost index and the non-fragile controller gains can be derived via adopting LMI Toolbox in Matlab to solve the corresponding convex optimization problem.

References

1. Chang, S.S., Peng, T.: Adaptive guaranteed cost control of systems with uncertain parameters. IEEE Trans. Autom. Control **17**(4), 474–483 (1972)
2. Pang, B., Liu, X., Jin, Q., Zhang, W.: Exponentially stable guaranteed cost control for continuous and discrete-time takagicsugeno fuzzy systems. Neurocomputing **205**, 210–221 (2016)
3. Ren, J., Zhang, Q.: Robust normalization and guaranteed cost control for a class of uncertain descriptor systems. Automatica **48**(8), 1693–1697 (2012)
4. Qiu, L., Yao, F., Xu, G., Li, S., Xu, B.: Output feedback guaranteed cost control for networked control systems with random packet dropouts and time delays in forward and feedback communication links. IEEE Trans. Autom. Sci. Eng. **13**(1), 284–295 (2016)
5. Lu, R., Cheng, H., Bai, J.: Fuzzy-model-based quantized guaranteed cost control of nonlinear networked systems. IEEE Trans. Fuzzy Syst. **23**(3), 567–575 (2015)
6. Han, C., Wu, L., Lam, H.K., Zeng, Q.: Nonfragile control with guaranteed cost of T-S fuzzy singular systems based on parallel distributed compensation. IEEE Trans. Fuzzy Syst. **22**(5), 1183–1196 (2014)
7. Li, X., Gao, H., Gu, K.: Delay-independent stability analysis of linear time-delay systems based on frequency discretization. Automatica **70**, 288–294 (2016)
8. Li, Q.-K., Lin, H.: Effects of mixed-modes on the stability analysis of switched time-varying delay systems. IEEE Trans. Autom. Control **61**(10), 3038–3044 (2016)
9. Takagi, T., Sugeno, M.: Fuzzy identification of systems and its applications to modeling and control. IEEE Trans. Syst. Man Cybern. **15**(1), 116–132 (1985)
10. Himavathi, S., Umamaheswari, B.: New membership functions for effective design and implementation of fuzzy systems. IEEE Trans. Syst. Man Cybern. Part A: Syst. Hum. **31**(6), 717–723 (2001)

11. Tao, J., Lu, R., Shi, P., Su, H., Wu, Z.-G.: Dissipativity-based reliable control for fuzzy Markov jump systems with actuator faults. IEEE Trans. Cybern. **47**(9), 2377–2388 (2017)
12. Zhang, L., Ning, Z., Shi, P.: Input-output approach to control for fuzzy Markov jump systems with time-varying delays and uncertain packet dropout rate. IEEE Trans. Cybern. **45**(11), 2449–2460 (2015)

Chapter 4
Quantized Control of Fuzzy Hidden MJSs

4.1 Introduction

The primary goal of this chapter is to investigate the GCC problem for nonlinear MJSs affected by quantization. Based on the HMM and the T–S fuzzy approach, we devote to designing an asynchronous controller, which can minimize the GCC performance index. Besides, the quantizer is also assumed to operate asynchronously with the plant, which is conditionally independent of the controller. The sector bound approach is used to handle quantization errors. By utilization of the Lyapunov function with some slack matrices, sufficient conditions are derived to ensure the stochastic stability of the closed-loop system with GCC performance. The solution to controller gains is given in the form of LMIs.

4.2 Preliminary Analysis

Consider the following T–S fuzzy MJSs:
Plant rule i: IF θ_{1k} is ζ_{i1}, θ_{2k} is ζ_{i2}, ..., and θ_{vk} is ζ_{iv}, THEN

$$x(k+1) = A_{\alpha_k i}x(k) + B_{\alpha_k i}u(k), \tag{4.1}$$

where θ_{jk} $(j \in \{1, 2, \ldots, v\})$ is the premise variable, and ζ_{ij} $(i \in \mathcal{R} = \{1, 2, \ldots, r\})$ is the fuzzy set. r is the total number of fuzzy rules. $x(k) \in R^{n_x}$ represents the state vector, and $u(k) \in R^{n_u}$ is the controlled input. $A_{\alpha_k i}$ and $B_{\alpha_k i}$ are given real matrices with appropriate dimensions. $\alpha_k \in \mathcal{L}$ $(\mathcal{L} = \{1, 2, \ldots, l\})$ is applied to describe the discrete-time Markov jump that is subject to the given transition probability matrix $\Pi = [\mu_{ab}]$ with

$$\Pr\{\alpha_{k+1} = b | \alpha_k = a\} = \mu_{ab} \tag{4.2}$$

and it is worth pointing out that $\mu_{ab} \in [0, 1]$ and $\sum_{b=1}^{l} \mu_{ab} = 1$.

S. Dong et al., *Control and Filtering of Fuzzy Systems with Switched Parameters*, Studies in Systems, Decision and Control 268,
https://doi.org/10.1007/978-3-030-35566-1_4

Via the T–S fuzzy approach, system (4.1) is inferred as

$$x(k+1) = A_{\alpha_k h}x(k) + B_{\alpha_k h}u(k), \tag{4.3}$$

where

$$A_{\alpha_k h} = \sum_{i=1}^{r} h_i(\theta_k)A_{\alpha_k i},\ B_{\alpha_k h} = \sum_{i=1}^{r} h_i(\theta_k)B_{\alpha_k i},$$

$$\theta_k = [\theta_{1k}, \theta_{2k}, \ldots, \theta_{vk}],\ h_i(\theta_k) = \frac{\prod_{j=1}^{v}\zeta_{ij}(\theta_{jk})}{\sum_{i=1}^{r}\prod_{j=1}^{v}\zeta_{ij}(\theta_{jk})}.$$

$\zeta_{ij}(\theta_{jk})$ stands for the grade of membership θ_{jk} in ζ_{ij}. It is assumed that $\prod_{j=1}^{v}\zeta_{ij}(\theta_{jk}) \geq 0$. Therefore, we obtain that $h_i(\theta_k) \geq 0$ and $\sum_{i=1}^{r} h_i(\theta_k) = 1$.

Based on the PDC technique, we devise the following asynchronous fuzzy controller:

Controller rule i: IF θ_{1k} is ζ_{i1}, θ_{2k} is ζ_{i2}, ..., and θ_{vk} is ζ_{iv}, THEN

$$u(k) = K_{\beta_k i}x(k), \tag{4.4}$$

where $K_{\beta_k i}$ is the controller gain to be designed. $\beta_k \in \mathcal{M}$ ($\mathcal{M} = \{1, 2, \ldots, m\}$) represents an HMM process and satisfies the given conditional probability matrix $\Upsilon = [\lambda_{ac}]$ with

$$\Pr\{\beta_k = c|\alpha_k = a\} = \lambda_{ac}, \tag{4.5}$$

where $\lambda_{ac} \in [0, 1]$ and $\sum_{c=1}^{m}\lambda_{ac} = 1$. Accordingly, the overall control law is

$$u(k) = K_{\beta_k h}x(k), \tag{4.6}$$

where $K_{\beta_k h} = \sum_{i=1}^{r} h_i(\theta_k)K_{\beta_k i}$.

Remark 4.1 Despite of extensive research about MJSs, it is often assumed that the designed controller (or filter) has the same mode as the original system [1, 2]. The problem of accessibility to original system modes gets scarce attention. Recently, the HMM, a stochastic process, has been introduced to function as a detector (or an observer) to obtain the original system modes, shown in (4.5). The detector-based (observer-based) formulation classifies the observation results into complete observation, cluster observation and no information in some published works [3–6]. In the chapter, we are more concerned with that whether the mode of the original fuzzy MJSs and the designed controller (or filter) is identical, which has been investigated in [7]. It should be noted that mode-independence ($\mathcal{M} = \{1\}$) and synchronization ($\Pr\{\beta_k = a|\alpha_k = a\} = \lambda_{aa} = 1$) can be seen as special cases of asynchronous phenomena.

Remark 4.2 Until now, according to our knowledge, there exist two kinds of asynchronous Markov principles, namely, the piecewise Markov jump and the HMM. The first one has been applied in [8, 9], which is in the control of the modes from both the original system and the designed filter/controller. By contrast, the latter directly depends on the present mode of the original system, which is a compact asynchronous mechanism.

In practice, since capacity of communication channels is limited, it is quite important to quantize $u(k)$ before being sent for reducing the quantities of data transmitted. The stochastic logarithmic quantizer is adopted, i.e.,

$$\mathcal{Q}_{\tau_k}(u) = [\mathcal{Q}_{\tau_k 1}(u_1), \mathcal{Q}_{\tau_k 2}(u_2), \ldots, \mathcal{Q}_{\tau_k n_u}(u_{n_u})]^T, \tag{4.7}$$

where $u = [u_1, u_2, \ldots, u_{n_u}]$. n_u is the number of used logarithmic quantizers, which is symmetric, i.e., $\mathcal{Q}_{\tau_k i}(u_i) = -\mathcal{Q}_{\tau_k i}(-u_i)$. $\tau_k \in \mathcal{N}$ ($\mathcal{N} = \{1, 2, \ldots, n\}$) is employed to describe the stochastic phenomenon of operational quantizers, represented by another HMM with the given conditional probability matrix $V = [\nu_{ad}]$ and

$$\Pr\{\tau_k = d | \alpha_k = a\} = \nu_{ad}, \tag{4.8}$$

where $\nu_{ad} \in [0, 1]$ and $\sum_{d=1}^{n} \nu_{ad} = 1$. For each quantizer in each working mode, the set of the logarithmic quantization levels are defined as

$$\begin{aligned} W_{\tau_k i} = \{\pm\omega^g_{\tau_k i} : \omega^g_{\tau_k i} = \rho^g_{\tau_k i}\omega^0_{\tau_k i}, g = \pm 1, \pm 2, \ldots\} \cup \{0\} \\ (0 < \rho_{\tau_k i} < 1,\ u_{i0} > 0), \end{aligned} \tag{4.9}$$

where $\rho_{\tau_k i}$ denotes the given quantization density of the ith quantizer when its mode τ_k works. g means the quantization level. $\rho^g_{\tau_k i}\omega^0_{\tau_k i}$ (or $\omega^g_{\tau_k i}$) is the output of the ith operational quantizer in mode τ_k at the quantization level g, and $\omega^0_{\tau_k i}$ is the initial quantization state, shown in Fig. 4.1. The stochastic quantizer function is

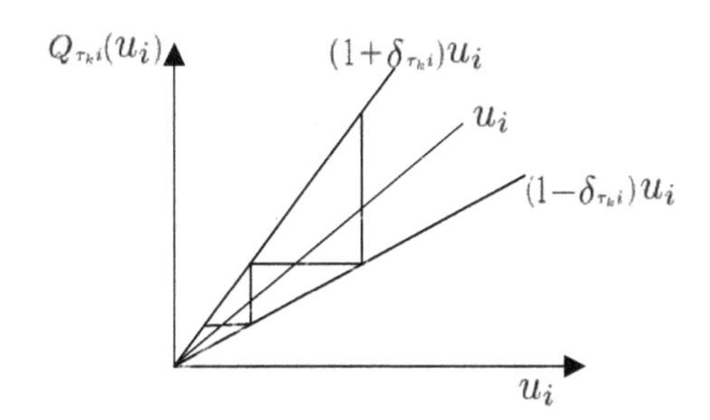

Fig. 4.1 Stochastic quantization model

$$Q_{\mathcal{T}_k i}(u_i) = \begin{cases} \omega^g_{\mathcal{T}_k i}, & \frac{\omega^g_{\mathcal{T}_k i}}{1+\delta_{\mathcal{T}_k i}} \le u_i \le \frac{\omega^g_{\mathcal{T}_k i}}{1-\delta_{\mathcal{T}_k i}}, \\ 0, & u_i = 0, \\ -Q_{\mathcal{T}_k i}(-u_i), & u_i < 0, \end{cases} \tag{4.10}$$

where $\delta_{\mathcal{T}_k i} = \frac{1-\rho_{\mathcal{T}_k i}}{1+\rho_{\mathcal{T}_k i}}$. It is evident that $\delta_{\mathcal{T}_k i} \in (0, 1)$.

We adopt the sector bound approach [10] to analyze quantization errors:

$$Q_{\mathcal{T}_k i}(u_i) = (1 + \Delta_{\mathcal{T}_k i}(u_i))u_i, \ |\Delta_{\mathcal{T}_k i}(u_i)| < \delta_{\mathcal{T}_k i}. \tag{4.11}$$

Defining $u_i = u_i(k)$, we get $Q_{\mathcal{T}_k i}(u_i(k)) = (1 + \Delta_{\mathcal{T}_k i}(u_i(k)))u_i(k)$. For notational simplicity, $\Delta_{\mathcal{T}_k i}(u_i(k))$ is expressed as $\Delta_{\mathcal{T}_k i}(k)$. As a result, $Q_{\mathcal{T}_k}(u(k))$ is deduced as

$$Q_{\mathcal{T}_k}(u(k)) = (I + \Delta_{\mathcal{T}_k}(k))u(k), \tag{4.12}$$

where $\Delta_{\mathcal{T}_k}(k) = \text{diag}\{\Delta_{\mathcal{T}_k 1}(k), \ldots, \Delta_{\mathcal{T}_k n_u}(k)\}$. From $|\Delta_{\mathcal{T}_k i}(k)| < \delta_{\mathcal{T}_k i} < 1$, it follows that $\Delta^T_{\mathcal{T}_k}(k)\Delta_{\mathcal{T}_k}(k) < \bar{\Delta}^T_{\mathcal{T}_k}\bar{\Delta}_{\mathcal{T}_k} < I$ $(\bar{\Delta}_{\mathcal{T}_k} = \text{diag}\{\delta_{\mathcal{T}_k 1}, \ldots, \delta_{\mathcal{T}_k n_u}\})$.

Remark 4.3 It is worth pointing out that literature concerning the quantization effect for MJSs often assumes that the quantizer is independent of the original system modes, for instance, [9, 11]. This will ignore some useful mode information and cannot better describe practical phenomena. In this chapter, a more general stochastic quantization model is constructed by adopting the HMM as well, shown in (4.8). Its function is similar to that of the designed controller (4.6) and it also covers two special cases: mode-independence and synchronization. However, the used stochastic variables β_k and $\mathcal{T}_k$ are conditionally independent, namely,

$$\begin{aligned} &\Pr\{\beta_k = c, \mathcal{T}_k = d | \alpha_k = a\} \\ &= \Pr\{\beta_k = c | \alpha_k = a\} \times \Pr\{\mathcal{T}_k = d | \alpha_k = a\} \\ &= \lambda_{ac}\nu_{ad}. \end{aligned} \tag{4.13}$$

For simplicity, $a, \ c, \ d$ and h_i are used to replace $\alpha_k, \ \beta_k, \ \mathcal{T}_k$ and $h_i(\theta_k)$ in the following.

Via incorporating (4.6) and (4.12) into (4.3), the closed-loop system is obtained:

$$x(k+1) = \bar{A}_{acdh}x(k) \tag{4.14}$$

where

$$\bar{A}_{acdh} = A_{ah} + B_{ah}(I + \Delta_d(k))K_{ch} = \sum_{i=1}^{r}\sum_{j=1}^{r} h_i h_j \bar{A}_{acdij},$$

$$\bar{A}_{acdij} = A_{ai} + B_{ai}(I + \Delta_d(k))K_{cj}.$$

We introduce the following GCC performance to design a guaranteed cost controller (4.6):

$$J_1 = \sum_{k=0}^{\infty} E\left\{x^T(k)G_1x(k) + u^T(k)G_2u(k)\right\}, \tag{4.15}$$

where matrices G_1 and G_2 are given and positive-definite. Considering quantization effects on $u(k)$ in (4.12) and the state feedback control (4.6), the GCC performance index is inferred as

$$J = \sum_{k=0}^{\infty} E\left\{x^T(k)(G_1 + \tilde{K}_{cdh}^T G_2 \tilde{K}_{cdh})x(k)\right\}, \tag{4.16}$$

where

$$\tilde{K}_{cdh} = (I + \Delta_d(k))K_{ch}.$$

The aim of our chapter is to devise the controller meeting the following conditions:
(1) The system (4.14) is stochastically stable;
(2) The GCC performance index (4.16) satisfies $J \leq J_{opt}$, where J_{opt} is the minimal upper bound of the GCC criterion.

4.3 Main Results

The first step of this section is to propose a sufficient condition ensuring the stability of closed-loop systems and find an upper bound for GCC performance, provided that controller parameters are given. Then, we focus on developing an approach to obtain controller (4.6).

Theorem 4.1 *If there exist matrices $U_{acdi} > 0,\ P_{ai} > 0$ and diagonal matrices $M_d > 0$ for any $a \in \mathcal{L},\ c \in \mathcal{M},\ d \in \mathcal{N}$ and $i, j, t \in \mathcal{R}$ satisfying*

$$\sum_{c=1}^{m}\sum_{d=1}^{n} \lambda_{ac}\nu_{ad}U_{acdi} < P_{ai}, \tag{4.17}$$

$$\Xi_{acdiit} < 0, \tag{4.18}$$

$$\Xi_{acdijt} + \Xi_{acdjit} < 0,\ i < j, \tag{4.19}$$

system (4.14) is stochastically stable and the GCC performance meets

$$J < x^T(0)P_{\alpha_0}(0)x(0), \tag{4.20}$$

where

$$\Xi_{acdijt} = \begin{bmatrix} -U_{acdi} & * & * & * & * & * \\ \Xi_{21} & \Xi_{22} & * & * & * & * \\ I & 0 & -G_1^{-1} & * & * & * \\ K_{cj} & 0 & 0 & -G_2^{-1} & * & * \\ 0 & \Xi_{52} & 0 & M_d\bar{\Delta}_d & -M_d & * \\ K_{cj} & 0 & 0 & 0 & 0 & -M_d \end{bmatrix},$$

$$\Xi_{21} = \begin{bmatrix} \sqrt{\mu_{a1}}P_{1t}(A_{ai} + B_{ai}K_{cj}) \\ \vdots \\ \sqrt{\mu_{al}}P_{lt}(A_{ai} + B_{ai}K_{cj}) \end{bmatrix},$$

$$\Xi_{22} = -diag\{P_{1t},\ P_{2t}, \ldots,\ P_{lt}\},$$

$$\Xi_{52} = M_d\bar{\Delta}_d B_{ai}^T \left[\sqrt{\mu_{a1}}P_{1t} \cdots \sqrt{\mu_{al}}P_{lt}\right].$$

Proof To begin with, define

$$\begin{aligned} &P_a(k) = P_{ah} = \sum_{i=1}^{r} h_i P_{ai},\ P_b(k+1) = P_{bh^+} = \sum_{t=1}^{r} h_t^+ P_{bt}, \\ &U_{acdh} = \sum_{i=1}^{r} h_i U_{acdi},\ h_t^+ = h_t(\theta_{k+1}),\ t \in \mathcal{R}. \end{aligned} \tag{4.21}$$

From (4.14), (4.17)–(4.19) and (4.21), it follows that

$$\sum_{c=1}^{m}\sum_{d=1}^{n} \lambda_{ac}\nu_{ad}U_{acdh} < P_{ah}, \tag{4.22}$$

and

$$\begin{aligned} \Xi_{hh^+} = &\sum_{t=1}^{r}\sum_{i=1}^{r}\sum_{j=1}^{r} h_t^+ h_i h_j \Xi_{acdijt} = \sum_{t=1}^{r} h_t^+ \left(\sum_{i=1}^{r} h_i^2 \Xi_{acdiit} \right. \\ &\left. + \sum_{i=1}^{r-1}\sum_{j=i+1}^{r} h_i h_j (\Xi_{acdijt} + \Xi_{acdjit}) \right) < 0, \end{aligned} \tag{4.23}$$

where

$$\Xi_{hh^+} = \begin{bmatrix} \Xi_{hh^+}^{11} & * & * \\ M_d\bar{\Delta}_d\Xi_{hh^+}^{21} & -M_d & * \\ \Xi_{hh^+}^{31} & 0 & -M_d \end{bmatrix},$$

$$\Xi_{hh^+}^{11} = \begin{bmatrix} -U_{acdh} & * & * & * \\ \Xi_{21}' & \Xi_{22}' & * & * \\ I & 0 & -G_1^{-1} & * \\ K_{ch} & 0 & 0 & -G_2^{-1} \end{bmatrix},$$

$$
\begin{aligned}
&\Xi_{hh^+}^{21} = \left[0\ \Xi_{52}'\ 0\ I\right],\ \Xi_{hh^+}^{31} = \left[K_{ch}\ 0\ 0\ 0\right],\\
&\Xi_{21}' = \begin{bmatrix} \sqrt{\mu_{a1}} P_{1h^+}(A_{ah} + B_{ah}K_{ch}) \\ \vdots \\ \sqrt{\mu_{al}} P_{lh^+}(A_{ah} + B_{ah}K_{ch}) \end{bmatrix},\\
&\Xi_{22}' = -\mathrm{diag}\{P_{1h^+},\ P_{2h^+}, \ldots,\ P_{lh^+}\},\\
&\Xi_{52}' = B_{ah}^T \left[\sqrt{\mu_{a1}} P_{1h^+} \cdots \sqrt{\mu_{al}} P_{lh^+}\right].
\end{aligned}
$$

Considering $M_d > 0$, by applying Schur Complement to (4.23), we have

$$
\Xi_{hh^+}^{11} + (\Xi_{hh^+}^{21})^T \bar{\Delta}_d M_d \bar{\Delta}_d \Xi_{hh^+}^{21} + (\Xi_{hh^+}^{31})^T M_d^{-1} \Xi_{hh^+}^{31} < 0. \tag{4.24}
$$

Recalling $\Delta_d(k) \leq \bar{\Delta}_d$ in (4.12), we obtain

$$
\Xi_{hh^+}^{11} + (\Xi_{hh^+}^{21})^T \Delta_d(k) M_d \Delta_d(k) \Xi_{hh^+}^{21} + (\Xi_{hh^+}^{31})^T M_d^{-1} \Xi_{hh^+}^{31} < 0. \tag{4.25}
$$

Further, it follows that

$$
\Xi_{hh^+}^{11} + (\Xi_{hh^+}^{21})^T \Delta_d(k) \Xi_{hh^+}^{31} + (\Xi_{hh^+}^{31})^T \Delta_d(k) \Xi_{hh^+}^{21} < 0, \tag{4.26}
$$

namely,

$$
\begin{bmatrix} -U_{acdh} & * & * & * \\ \Xi_{21}'' & \Xi_{22}' & * & * \\ I & 0 & -G_1^{-1} & * \\ \tilde{K}_{cdh} & 0 & 0 & -G_2^{-1} \end{bmatrix} < 0 \tag{4.27}
$$

with

$$
\Xi_{21}'' = \begin{bmatrix} \sqrt{\mu_{a1}} P_{1h^+} \bar{A}_{acdh} \\ \vdots \\ \sqrt{\mu_{al}} P_{lh^+} \bar{A}_{acdh} \end{bmatrix},\ \tilde{K}_{cdh} = (I + \Delta_d(k))K_{ch}. \tag{4.28}
$$

Adopting Schur Complement to (4.27), we have

$$
G_1 + \tilde{K}_{cdh}^T G_2 \tilde{K}_{cdh} - U_{acdh} + \left(\sum_{b=1}^{l} \mu_{ab} \bar{A}_{acdh}^T P_{bh^+} \bar{A}_{acdh}\right) < 0. \tag{4.29}
$$

Based on (4.22), it is easy to observe that

$$
G_1 + \sum_{c=1}^{m} \sum_{d=1}^{n} \lambda_{ac} \nu_{ad} \tilde{K}_{cdh}^T G_2 \tilde{K}_{cdh} + \Theta < 0 \tag{4.30}
$$

with

$$\Theta = \sum_{c=1}^{m} \sum_{d=1}^{n} \lambda_{ac} \nu_{ad} \left(\sum_{b=1}^{l} \mu_{ab} \bar{A}_{acdh}^T P_{bh^+} \bar{A}_{acdh} \right) - P_{ah}.$$

From $G_1 > 0$ and $G_2 > 0$, it follows that $\Theta < 0$.

The candidate Lyapunov function is

$$V(k) = x^T(k) P_{ah} x(k). \tag{4.31}$$

Computing the forward difference of $V(k)$ and then taking expectation under $\alpha_k = a$, we have

$$\begin{aligned} E\{\Delta V(k)\} &= E\{x^T(k+1) P_{bh^+} x_(k+1)\} - E\{x^T(k) P_{ah} x(k)\} \\ &= E\left\{ x^T(k) \left(\bar{A}_{acdh}^T \sum_{b=1}^{l} \mu_{ab} P_{bh^+} \bar{A}_{acdh} - P_{ah} \right) x(k) \right\} \\ &= x^T(k) \Theta x(k) < 0, \end{aligned} \tag{4.32}$$

which means that system (4.14) is stochastically stable.

On the other hand,

$$\begin{aligned} &E\{\Delta V(k) + x^T(k)(G_1 + \tilde{K}_{cdh}^T G_2 \tilde{K}_{cdh}) x(k)\} \\ &= x^T(k) \left(G_1 + \sum_{c=1}^{m} \sum_{d=1}^{n} \lambda_{ac} \nu_{ad} \tilde{K}_{cdh}^T G_2 \tilde{K}_{cdh} + \Theta \right) x(k). \end{aligned} \tag{4.33}$$

According to the inequality (4.30), summing up from $k = 0$ to ∞ and recalling (4.16), we obtain

$$J < E\{V(0)\} - E\{V(\infty)\} < E\{V(0)\}. \tag{4.34}$$

In conclusion, this proves the Theorem 4.1 can guarantee system (4.14) stability with the bounded cost.

Now, based on the sufficient condition in Theorem 4.1 and the convex linearization approach, we focus on solving the GCC problem in the following.

Theorem 4.2 *If there exist matrices $\bar{P}_{ai} > 0$, $\bar{U}_{acdi} > 0$, Y_c, diagonal matrices $M_d > 0$ and $\bar{K}_{cj}$ for any $a \in \mathcal{L}$, $c \in \mathcal{M}$, $d \in \mathcal{N}$ and $i, j, t \in \mathcal{R}$ satisfying*

$$\begin{bmatrix} -\mathscr{U}_i & * \\ \mathscr{P}_{ai} & -\bar{P}_{ai} \end{bmatrix} < 0, \tag{4.35}$$

$$\Gamma_{acdiit} < 0, \tag{4.36}$$

$$\Gamma_{acdijt} + \Gamma_{acdjit} < 0, \ j > i, \tag{4.37}$$

system (4.14) is stochastically stable and GCC performance meets

$$J < x^T(0)\bar{P}_{\alpha_0}^{-1}(0)x(0). \tag{4.38}$$

The controller parameters can be solved by

$$K_{cj} = \bar{K}_{cj}Y_c^{-1}, \tag{4.39}$$

where

$$\begin{aligned}
&\mathscr{P}_{ai} = \bar{P}_{ai}\left[\sqrt{\phi_{a11}}I \ \sqrt{\phi_{a12}}I \ \cdots \ \sqrt{\phi_{amn}}I\right],\\
&\mathscr{U}_i = diag\{\bar{U}_{a11i}, \ \bar{U}_{a12i}, \ldots, \ \bar{U}_{amni}\}, \ \phi_{acd} = \lambda_{ac}\nu_{ad},\\
&\Gamma_{acdijt} = \begin{bmatrix} \Gamma_{11} & * & * & * & * & * \\ \Gamma_{21} & \Gamma_{22} & * & * & * & * \\ Y_c & 0 & -G_1^{-1} & * & * & * \\ \bar{K}_{cj} & 0 & 0 & -G_2^{-1} & * & * \\ 0 & \Gamma_{52} & 0 & M_d\bar{\Delta}_d & -M_d & * \\ \bar{K}_{cj} & 0 & 0 & 0 & 0 & -M_d \end{bmatrix},\\
&\Gamma_{11} = \bar{U}_{acdi} - Y_c - Y_c^T, \ \Gamma_{22} = -diag\{\bar{P}_{1t}, \ \bar{P}_{2t}, \ldots, \ \bar{P}_{lt}\},\\
&\Gamma_{21} = \begin{bmatrix} \sqrt{\mu_{a1}}(A_{ai}Y_c + B_{ai}\bar{K}_{cj}) \\ \vdots \\ \sqrt{\mu_{al}}(A_{ai}Y_c + B_{ai}\bar{K}_{cj}) \end{bmatrix},\\
&\Gamma_{52} = M_d\bar{\Delta}_d\left[\sqrt{\mu_{a1}}B_{ai}^T \ \cdots \ \sqrt{\mu_{al}}B_{ai}^T\right].
\end{aligned}$$

Proof Define

$$\begin{aligned}
&\bar{P}_{at} = P_{at}^{-1}, \ \bar{U}_{acdi} = U_{acdi}^{-1}, \mathscr{H} = \text{diag}\{I, -\Gamma_{22}, I, I, I, I\},\\
&\mathscr{Y} = \text{diag}\{Y_c, I, I, I, I, I\}.
\end{aligned} \tag{4.40}$$

By applying Schur Complement to (4.35), we have

$$\mathscr{P}_{ai}\mathscr{U}_i^{-1}\mathscr{P}_{ai}^T < \bar{P}_{ai}. \tag{4.41}$$

It is obvious that (4.41) is equivalent to (4.17).

Carrying out a congruence transformation to (4.18), we have

$$\mathscr{Y}^T\mathscr{H}\Xi_{acdiit}\mathscr{H}\mathscr{Y} < 0, \tag{4.42}$$

namely,

$$\Gamma'_{acdiit} = \begin{bmatrix} \Gamma'_{11} & * & * & * & * & * \\ \Gamma_{21} & \Gamma_{22} & * & * & * & * \\ Y_c & 0 & -G_1^{-1} & * & * & * \\ \bar{K}_{ci} & 0 & 0 & -G_2^{-1} & * & * \\ 0 & \Gamma_{52} & 0 & M_d\Delta_d & -M_d & * \\ \bar{K}_{ci} & 0 & 0 & 0 & 0 & -M_d \end{bmatrix} < 0 \tag{4.43}$$

with $\bar{K}_{ci} = K_{ci}Y_c$, $\Gamma'_{11} = -Y_c^T U_{acdi} Y_c$. From $(U_{acdi}^{-1} - Y_c^T)U_{acdi}(U_{acdi}^{-1} - Y_c) > 0$, it follows that

$$-Y_c^T U_{acdi} Y_c < \bar{U}_{acdi} - Y_c - Y_c^T. \tag{4.44}$$

Hence, $\Gamma_{acdiit} < 0 \Rightarrow \Gamma'_{acdiit} < 0$, which means that we can deduce (4.18) from (4.36). By the same way, (4.37) can guarantee that (4.19) holds. Accordingly, the controller parameters can be devised by (4.39). This completes the proof.

Remark 4.4 Compared with common Lyapunov function matrices, P_{ai} of our chapter is fuzzy-basis-dependent and mode-dependent, which is less conservative [12]. The slack matrix U_{acdi} is adopted for eliminating effects of conditional probabilities λ_{ac} and ν_{ad}, as shown in (4.29)–(4.30). Furthermore, we employ the slack matrix Y_c to separate P_{ai} from K_{cj}, transforming nonlinear matrix inequalities in Theorem 4.1 into LMIs in Theorem 4.2. Moreover, the synchronous controller and mode-independent controller can be obtained by assuming $\lambda_{aa} = 1$, $\mathcal{L} = \mathcal{M}$, $\lambda_{a1} = 1$ and $m = 1$, respectively.

Theorem 4.2 provides an approach to design an asynchronous fuzzy controller. However, the bounded cost in (4.20) (or (4.38)) depends on $P_{\alpha_0}(0)$ (or $\bar{P}_{\alpha_0}(0)$), the initial jump mode α_0 and the initial state $x(0)$. To cope with such situations, the following theorems are put forward.

Theorem 4.3 *If there exist matrices $\bar{P}_{ai} > 0$, $\bar{U}_{acdi} > 0$, Y_c, diagonal matrices $M_d > 0$, $\bar{K}_{cj}$ and scalar $\gamma > 0$ for any $a \in \mathcal{L}$, $c \in \mathcal{M}$, $d \in \mathcal{N}$ and $i, j, t \in \mathcal{R}$ such that the following optimization problem has a solution:*

$$\begin{aligned} &\min \quad \gamma \\ &s.t. \quad (a)\ LMIs\ (4.35),\ (4.36)\ and\ (4.37), \\ &\qquad (b) \begin{bmatrix} \gamma I & * \\ I & \bar{P}_{ai} \end{bmatrix} > 0, \end{aligned}$$

system (4.14) is stochastically stable with the controller gain (4.39) and the minimal cost value $J_{opt} = \gamma x^T(0)x(0)$.

Proof It follows that

$$
\begin{aligned}
x^T(0)P_{\alpha_0}(0)x(0) &= \sum_{i=1}^{r} h_i x^T(0)P_{\alpha_0 i}x(0) \\
&\le \sum_{i=1}^{r} h_i \lambda_{max}(P_{\alpha_0 i})x^T(0)x(0) \\
&\le \lambda_{max}(P_{\alpha_0 i})x^T(0)x(0).
\end{aligned}
\tag{4.45}
$$

From (b), it follows that $\bar{P}_{ai}^{-1} < \gamma I$. Furthermore, we obtain that $\bar{P}_{\alpha_0 i}^{-1} < \gamma I$. Hence, $\exists\ \gamma,\ \lambda_{max}(P_{\alpha_0 i}) \le \gamma$ holds, which suggests that minimizing γ can guarantee the optimal upper-bounded GCC value.

If the initial state $x(0)$ is uncertain but bounded, subject to $x(0)^T x(0) \le w^2$ $(w > 0)$, we obtain another technique.

Theorem 4.4 *If there exist matrices* $\bar{P}_{ai} > 0$, $\bar{U}_{acdi} > 0$, Y_c, *diagonal matrices* $M_d > 0$, $\bar{K}_{cj}$ *and scalar* $\varpi > 0$ *for any* $a \in \mathcal{L}$, $c \in \mathcal{M}$, $d \in \mathcal{N}$ *and* $i, j, t \in \mathcal{R}$ *such that the following optimization problem has a solution:*

$$
\begin{aligned}
&min \quad \varpi \\
&s.t. \quad (a)\ LMIs\ (4.35),\ (4.36)\ and\ (4.37), \\
&\qquad (b) \begin{bmatrix} \varpi I & * \\ wI & \bar{P}_{ai} \end{bmatrix} > 0,
\end{aligned}
$$

system (4.14) is stochastically stable with the controller gain (4.39) and the minimal cost value $J_{opt} = \varpi$.

Proof We have that

$$
\begin{aligned}
x^T(0)P_{\alpha_0}(0)x(0) &= \sum_{i=1}^{r} h_i x^T(0)P_{\alpha_0 i}x(0) \\
&\le \sum_{i=1}^{r} h_i \lambda_{max}(P_{\alpha_0 i})x^T(0)x(0) \\
&\le \lambda_{max}(P_{\alpha_0 i})x^T(0)x(0) \\
&\le \lambda_{max}(P_{\alpha_0 i})w^2.
\end{aligned}
\tag{4.46}
$$

The remaining proof is similar to that in Theorem 4.3. Therefore, we omit it.

Remark 4.5 Theorem 4.3 presents the optimization method for the GCC problem with the given initial state, which has no relation with the jump mode α_k. On the other hand, the optimization technique in Theorem 4.4 is independent of α_k and

$x(0)$, which considers a special phenomenon that the initial state has an upper bound without a certain value. In both cases, we can adopt LMI Toolbox in Matlab to obtain the optimal guaranteed cost controller and the smallest upper-bounded cost value.

4.4 Illustrative Example

In this section, two examples are provided to show the effectiveness and correctness of our proposed approach.

Example 1 Consider the following T–S fuzzy model for Hénon system without time-varying delays [13]:
Plant rule 1: IF $x_1(k)$ is $-m_{\alpha_k}$, THEN

$$x(k+1) = A_{\alpha_k 1}x(k) + B_{\alpha_k 1}u(k),$$

Plant rule 2: IF $x_1(k)$ is m_{α_k}, THEN

$$x(k+1) = A_{\alpha_k 2}x(k) + B_{\alpha_k 2}u(k),$$

where

$$\begin{aligned} A_{\alpha_k 1} &= \begin{bmatrix} m_{\alpha_k} & 0.3 \\ 1 & 0 \end{bmatrix}, \ A_{\alpha_k 2} = \begin{bmatrix} -m_{\alpha_k} & 0.3 \\ 1 & 0.1 \end{bmatrix}, \\ B_{\alpha_k 1} &= B_{\alpha_k 2} = \begin{bmatrix} 1 \\ 0 \end{bmatrix}, \ h_1(x_1(k)) = \frac{1}{2}\left(1 - \frac{x_1(k)}{m_{\alpha_k}}\right), \\ h_2(x_1(k)) &= 1 - h_1(x_1(k)), \ |x_1(k)| \le m_{\alpha_k}. \end{aligned}$$

The parameter m_{α_k} is assumed to be in the control of the Markov jump variable α_k, subject to

$$\Pi = \begin{bmatrix} 0.2 & 0.8 \\ 0.4 & 0.6 \end{bmatrix},$$

$m_1 = 0.9$, and $m_2 = 1$.

In this example, we assume that 1 quantizer works and has 3 jump modes with quantization densities: $\rho_{11} = 1, \ \rho_{21} = 0.8, \ \rho_{31} = 0.9$. By computation, we obtain that $\bar{\Delta}_d$ $(d = \{1, 2, 3\})$ in (4.10) are

$$\bar{\Delta}_1 = 0, \ \bar{\Delta}_2 = 0.1111, \ \bar{\Delta}_3 = 0.0526.$$

And we assume quantization errors $\Delta_i(k)$ in (4.12) to be

$$\Delta_1(k) = 0, \ \Delta_2(k) = 0.1111\sin(k), \ \Delta_3(k) = 0.0526\sin(k).$$

The jumps of the controller (4.6) and quantizer (4.12) are subject to the following probability matrices:

$$\Upsilon = \begin{bmatrix} 0.2 \ 0.5 \ 0.3 \\ 0.4 \ 0.4 \ 0.2 \end{bmatrix}, \ V = \begin{bmatrix} 0.3 \ 0.3 \ 0.4 \\ 0.2 \ 0.4 \ 0.4 \end{bmatrix}.$$

The GCC matrices G_1 and G_2 are supposed to be $G_1 = \text{diag}\{1, \ 1\}, \ G_2 = 0.25$.

Via solving LMIs in Theorem 4.3, we obtain the minimal γ with $\gamma = 2.2858$ and asynchronous controller gains as follows

$$\begin{aligned} K_{11} &= \begin{bmatrix} -0.9101 & -0.2761 \end{bmatrix}, \ K_{12} = \begin{bmatrix} 0.8735 & -0.2731 \end{bmatrix}, \\ K_{21} &= \begin{bmatrix} -0.9074 & -0.2714 \end{bmatrix}, \ K_{22} = \begin{bmatrix} 0.8594 & -0.2736 \end{bmatrix}, \\ K_{31} &= \begin{bmatrix} -0.9061 & -0.2705 \end{bmatrix}, \ K_{32} = \begin{bmatrix} 0.8559 & -0.2737 \end{bmatrix}. \end{aligned}$$

With the initial state $x(0) = \begin{bmatrix} 0.4 & -0.4 \end{bmatrix}^T$, Fig. 4.2 shows that the trajectories of the closed-system states converge to zero while Fig. 4.3 represents the curve of the quantized input, which demonstrate the effectiveness and applicability of our developed method.

Example 2 The single-link robot arm system in Chap. 3 is considered without time-varying delays, which is modelled as

$$\begin{cases} \dot{x}_1(t) = x_2(t), \\ \dot{x}_2(t) = -\dfrac{glM}{J}\sin(x_1(t)) - \dfrac{R}{J}x_2(t) + \dfrac{1}{J}u(t), \end{cases} \tag{4.47}$$

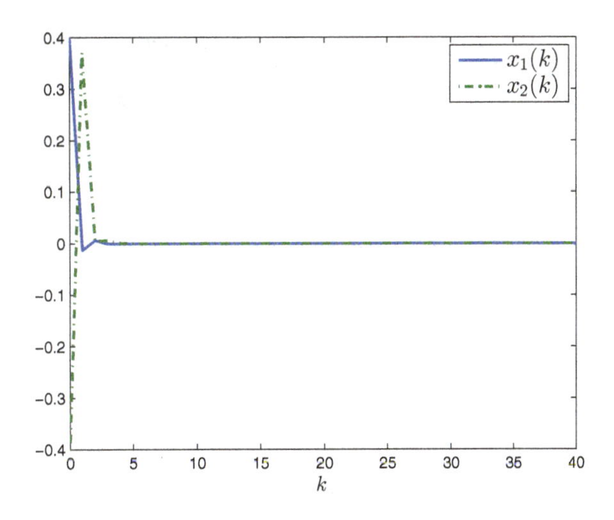

Fig. 4.2 State responses of the closed-loop system (4.14) in Example 1

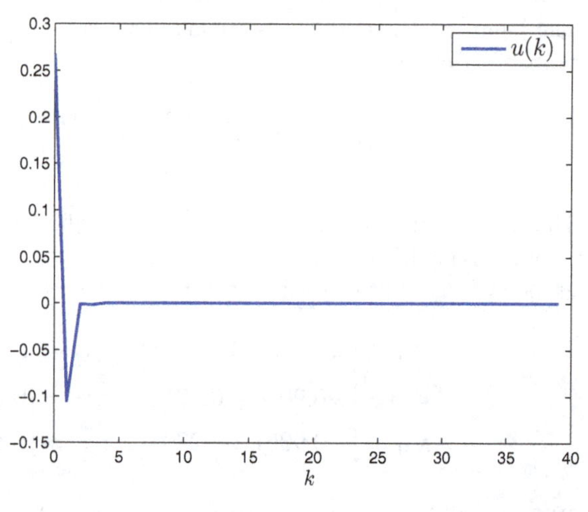

Fig. 4.3 Quantized controller input $u(k)$ in Example 1

Table 4.1 Values M_{α_k} and J_{α_k} for different modes α_k

α_k	1	2
M_{α_k}	1	3
J_{α_k}	1	3

where $l,\ g,\ M,\ J,$ and R are the robot arm length, the gravity acceleration, the payload mass, the inertia moment and the viscous friction coefficient, respectively.

With the sampling period $T = 0.1$, the first-order Euler Approximation technique is applied to discretize the system (4.47). We assume that the payload mass M and the inertia moment J have two modes, shown in Table 4.1. α_k is subject to the following transition probability matrix

$$\Pi = \begin{bmatrix} 0.3 & 0.7 \\ 0.7 & 0.3 \end{bmatrix}.$$

Then based on T–S fuzzy inference approach, the fuzzy Markov jump model is inferred as

Plant rule 1: IF $x_1(k)$ is about 0 rad, THEN

$$x(k+1) = A_{\alpha_k 1}x(k) + B_{\alpha_k 1}u(k),$$

Plant rule 2: IF $x_1(k)$ is about $\pm\pi$ rad, THEN

$$x(k+1) = A_{\alpha_k 2}x(k) + B_{\alpha_k 2}u(k),$$

where

$$A_{\alpha_k 1} = \begin{bmatrix} 1 & T \\ -\frac{TglM_{\alpha_k}}{J_{\alpha_k}} & 1 - \frac{TR}{J_{\alpha_k}} \end{bmatrix}, \ B_{\alpha_k 1} = \begin{bmatrix} 0 \\ \frac{T}{J_{\alpha_k}} \end{bmatrix},$$

$$A_{\alpha_k 2} = \begin{bmatrix} 1 & T \\ -\frac{\beta TglM_{\alpha_k}}{J_{\alpha_k}} & 1 - \frac{TR}{J_{\alpha_k}} \end{bmatrix}, \ B_{\alpha_k 2} = \begin{bmatrix} 0 \\ \frac{T}{J_{\alpha_k}} \end{bmatrix},$$

$$h_1(x_1(k)) = \begin{cases} \dfrac{\sin(x_1(k)) - \beta x_1(k)}{(1-\beta)x_1(k)}, & x_1(k) \neq 0, \\ 1, & x_1(k) = 0, \end{cases}$$

$$h_2(x_1(k)) = 1 - h_2(x_1(k)), \ \beta = 10^{-2}/\pi.$$

The conditional transition probability matrices Υ and V are

$$\Upsilon = \begin{bmatrix} 0.5 \ 0.5 \\ 0.6 \ 0.4 \end{bmatrix}, \ V = \begin{bmatrix} 0.3 \ 0.7 \\ 0.4 \ 0.6 \end{bmatrix}.$$

In this example, there is one quantizer working in 2 modes and quantization densities (4.10) are assumed to be $\rho_{11} = 0.6, \ \rho_{21} = 0.9$. Accordingly, we obtain $\bar{\Delta}_d$ $(d = \{1, 2\})$ in (4.10):

$$\bar{\Delta}_1 = 0.25, \ \bar{\Delta}_2 = 0.0526,$$

and we assume quantization errors $\Delta_i(k)$ in (4.12) to be

$$\Delta_1(k) = 0.25\cos(k), \ \Delta_2(k) = 0.0526\sin(k).$$

The GCC matrices G_1 and G_2 are

$$G_1 = \begin{bmatrix} 0.4 & 0 \\ 0 & 0.4 \end{bmatrix}, G_2 = 0.5.$$

Solving the optimization problem in Theorem 4.3 by LMI Toolbox, we obtain the asynchronous controller gains as follows:

$$K_{11} = \begin{bmatrix} -0.3309 & -1.4932 \end{bmatrix}, \ K_{12} = \begin{bmatrix} -1.4737 & -1.5810 \end{bmatrix},$$
$$K_{21} = \begin{bmatrix} -0.3172 & -1.5819 \end{bmatrix}, \ K_{22} = \begin{bmatrix} -1.5930 & -1.6845 \end{bmatrix},$$

and the minimal γ is 60.4062. By simulating with the initial state $x(0) = \begin{bmatrix} 0.4\pi & -0.4\pi \end{bmatrix}^T$, it is evident from Fig. 4.4 that states approach to zero under the control input in Fig. 4.5, which means that our proposed approach is valid.

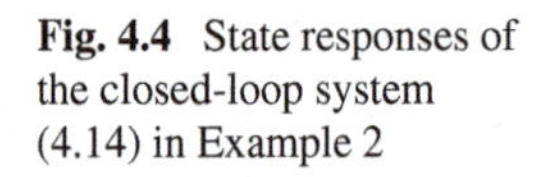

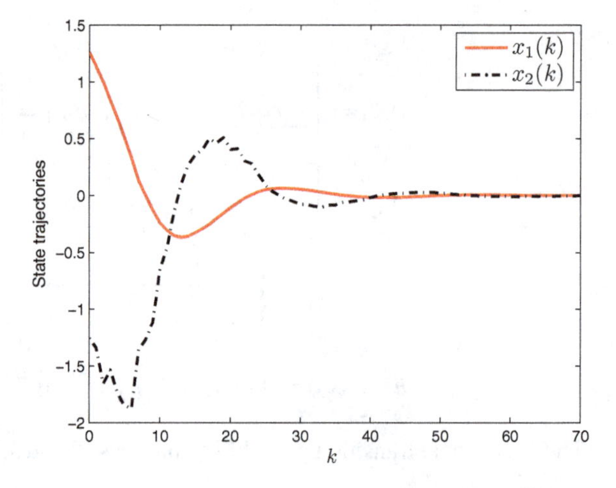

Fig. 4.4 State responses of the closed-loop system (4.14) in Example 2

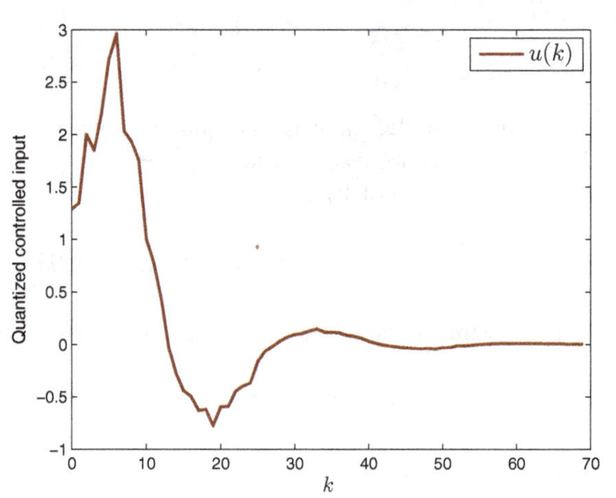

Fig. 4.5 Quantized controller input $u(k)$ in Example 2

In the following, we will analyze how the optimal GCC γ^* changes on the whole conditional probability space, i.e., Υ of the designed asynchronous controller and V of the stochastic quantizer, respectively.
(1) Relationship of γ^* and Υ: firstly, we let

$$\Upsilon = \begin{bmatrix} \lambda_{11} & 1-\lambda_{11} \\ \lambda_{21} & 1-\lambda_{21} \end{bmatrix}, \ V = \begin{bmatrix} 0.3 & 0.7 \\ 0.4 & 0.6 \end{bmatrix}.$$

Figure 4.6 shows the variation of r^* on the whole space and we can clearly find that when $(\lambda_{11}, \lambda_{21}) = (1, 0)$ or $(0, 1)$, γ^* is minimal with 41.1808. The former $(\lambda_{11}, \lambda_{21}) = (1, 0)$ means that the filter is synchronous with the plant while the latter represents the highest asynchronization situation. Figure 4.7 provides the two special cases: when $\lambda_{11} = \lambda_{21}$, r^* remains unchanged with 60.6135, which is the peak

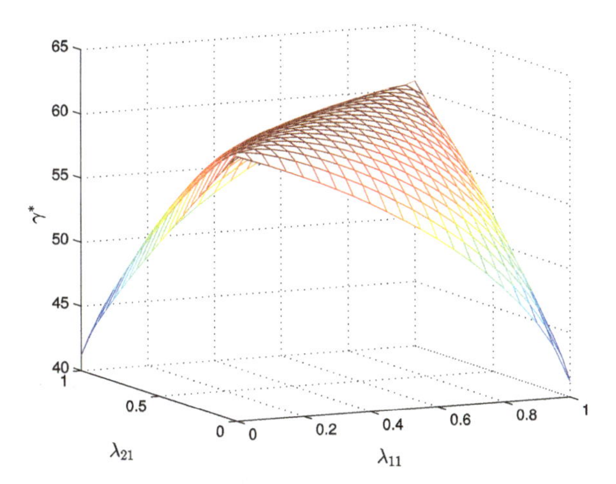

Fig. 4.6 Optimal GCC γ^* with Υ

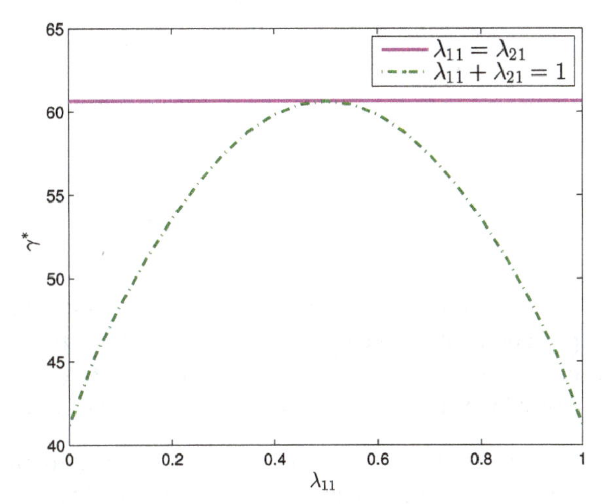

Fig. 4.7 Special cases of optimal GCC γ^* with Υ

value of γ^*; when $\lambda_{11} + \lambda_{21} = 1$, the curve is symmetric. In [0, 0.5], γ^* increases and it reaches the maximum 60.6135 at $\lambda_{11} = 0.5$. Then, it drops to the lowest point 41.1808 at $\lambda_{11} = 1$.

(2) Relationship of γ^* and V: let

$$\Upsilon = \begin{bmatrix} 0.5 & 0.5 \\ 0.6 & 0.4 \end{bmatrix}, \quad V = \begin{bmatrix} 0.5 & 0.5 \\ 0.5 & 0.5 \end{bmatrix}.$$

We assume that the quantizer has the same quantization densities in two modes, namely $\rho = \rho_{11} = \rho_{21}$. Via changing ρ, we find from Table 4.2 that the bigger ρ is, the smaller γ^* is, which means the GCC performance becomes better with higher quantization density. Furthermore, we focus on analyzing γ^* on the whole probability space of quantizer, and further suppose that

Table 4.2 Optimal GCC $\gamma*$ for different quantization density ρ

ρ	0.2	0.4	0.6	0.8	1.0
$\gamma*$	298.2894	98.4071	69.8852	59.2554	54.0138

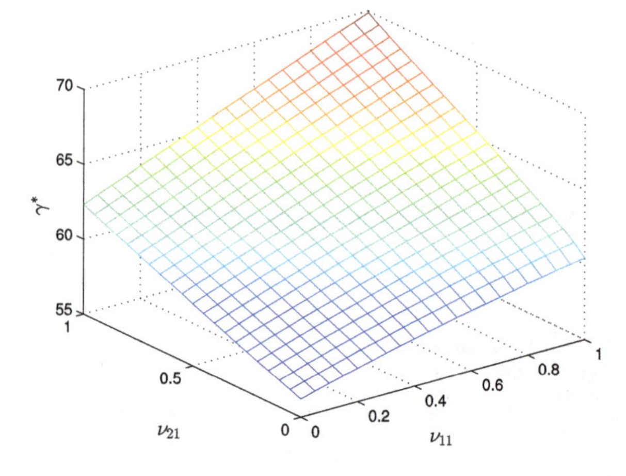

Fig. 4.8 Optimal GCC γ^* with V

$$\rho_{11} = 0.6, \ \rho_{21} = 0.9, \ V = \begin{bmatrix} \nu_{11} & 1 - \nu_{11} \\ \nu_{21} & 1 - \nu_{21} \end{bmatrix}.$$

Figure 4.8 presents that the higher ν_{11} or ν_{21} is, the bigger γ^* is. It implies that the probability of the quantizer running in 1st mode turns bigger while the quantizer of the 1st mode has lower quantization density. As a result, the system performance is worse (γ^* is larger), which accords with the quantizer feature.

Remark 4.6 From Figs. 4.6 and 4.7, we find that for the case $\lambda_{11} = \lambda_{21}$, the optimal value of GCC performance is independent of original system modes. However, we cannot say that obtained optimal GCC controller parameters have nothing to do with original system modes because the controller parameters are not always the same during simulation for this case. On the other hand, from Figs. 4.6 and 4.8, it is easy to observe that conditional transition probability matrices Υ and V on the whole space have different forms of changes though the designed controller and the quantizer have adopted the same asynchronous theory-HMM. This is mainly because of the intrinsic characteristic of quantizers: the more precise the quantizer is (its density is bigger), the better GCC performance we obtain (the value of γ^* is smaller), which coincides with simulation results in Table 4.2 and Fig. 4.8.

4.5 Conclusion

In this chapter, we have investigated the GCC problem for nonlinear MJSs modeled by the T–S fuzzy method. A more complex and general system is considered, where the quantizer and the designed controller run asynchronously with the plant, respectively. Moreover, via the Lyapunov function technique and the slack matrix approach, we have proposed three GCC methods for different phenomena in terms of LMIs. The optimal fuzzy controller parameters and the minimal upper-bounded cost can be achieved by solving a convex optimization problem. The simulation outcomes have demonstrated the feasibility of the theoretic results obtained.

References

1. Zhu, S., Han, Q.-L., Zhang, C.: l_1-gain performance analysis and positive filter design for positive discrete-time Markov jump linear systems: a linear programming approach. Automatica **50**(8), 2098–2107 (2014)
2. Gonzaga, C.A.C., Costa, O.L.V.: Stochastic stabilization and induced l_2-gain for discrete-time Markov jump Lur'e systems with control saturation. Automatica **50**(9), 2397–2404 (2014)
3. de Oliveira, A., Costa, O.: H_2-filtering for discrete-time hidden Markov jump systems. Int. J. Control **90**(3), 599–615 (2017)
4. Graciani Rodrigues, C., Todorov, M.G., Fragoso, M.D.: H_∞ control of continuous-time Markov jump linear systems with detector-based mode information. Int. J. Control **90**(10), 2178–2196 (2017)
5. do Valle Costa, O.L., Fragoso, M.D., Todorov, M.G.: A detector-based approach for the H_2 control of Markov jump linear systems with partial information. IEEE Trans. Autom. Control **60**(5), 1219–1234 (2015)
6. Stadtmann, F., Costa, O.: H_2-control of continuous-time hidden Markov jump linear systems. IEEE Trans. Autom. Control **62**(8), 4031–4037 (2017)
7. Wu, Z.-G., Shi, P., Shu, Z., Su, H., Lu, R.: Passivity-based asynchronous control for Markov jump systems. IEEE Trans. Autom. Control **62**(4), 2020–2025 (2017)
8. Wu, Z.-G., Shi, P., Su, H., Lu, R.: Asynchronous l_2-l_∞ filtering for discrete-time stochastic Markov jump systems with randomly occurred sensor nonlinearities. Automatica **50**(5), 180–186 (2014)
9. Zhang, L., Zhu, Y., Shi, P., Zhao, Y.: Resilient asynchronous H_∞ filtering for Markov jump neural networks with unideal measurements and multiplicative noises. IEEE Trans. Cybern. **45**(12), 2840–2852 (2015)
10. Fu, M., Xie, L.: The sector bound approach to quantized feedback control. IEEE Trans. Autom. Control **50**(11), 1698–1711 (2005)
11. Tao, J., Lu, R., Su, H., Shi, P., Wu, Z.-G.: Asynchronous filtering of nonlinear Markov jump systems with randomly occurred quantization via T-S fuzzy models. IEEE Trans. Fuzzy Syst. **26**(4), 1866–1877 (2018)
12. Wu, Z.-G., Dong, S., Su, H., Li, C.: Asynchronous dissipative control for fuzzy Markov jump systems. IEEE Trans. Cybern. **48**(8), 2426–2436 (2018)
13. Gao, H., Liu, X., Lam, J.: Stability analysis and stabilization for discrete-time fuzzy systems with time-varying delay. IEEE Trans. Syst. Man Cybern. Part B (Cybern.) **39**(2), 306–317 (2009)

Chapter 5
Asynchronous Control of Fuzzy MJSs Subject to Strict Dissipativity

5.1 Introduction

This chapter is concerned with the asynchronous dissipative control design problems for fuzzy MJSs in continuous-time and discrete-time domains, respectively. Fuzzy asynchronous controllers are constructed via applying the PDC approach and the HMM. We apply the Lyapunov function to achieve sufficient conditions, which ensure the stochastic stability of the closed-loop system with strict dissipativity performance. Moreover, the desired dissipative controller parameters can be obtained via Matlab Toolbox.

5.2 Preliminary Analysis of Continuous-Time Systems

Consider the following continuous-time T–S fuzzy MJSs:
Plant rule i: IF $\zeta_1(t)$ is ξ_{i1}, $\zeta_2(t)$ is ξ_{i2}, ..., and $\zeta_p(t)$ is ξ_{ip}, THEN

$$\begin{cases} \dot{x}(t) = A_{r(t)i}x(t) + B_{1r(t)i}u(t) + D_{1r(t)i}w(t), \\ z(t) = C_{r(t)i}x(t) + B_{2r(t)i}u(t) + D_{2r(t)i}w(t), \end{cases} \tag{5.1}$$

where $x(t) \in R^n$, $u(t) \in R^m$, $z(t) \in R^q$, and $w(t) \in R^a$ are the state vector, the control input, the controlled output and the external disturbance subject to $l_2[0, +\infty)$, respectively. These known system matrices $A_{r(t)i}$, $B_{1r(t)i}$, $B_{2r(t)i}$, $C_{r(t)i}$, $D_{1r(t)i}$, and $D_{2r(t)i}$ have appropriate dimensions. System (5.1) has r ($i \in \mathcal{R} = \{1, 2, \ldots, r\}$) fuzzy rules and i means the ith rule. $\zeta_j(t)$ ($j \in \{1, 2, \ldots, p\}$) is the premise variable. ξ_{ij} is the fuzzy set. The variable $r(t)$ ($r(t) \in \mathcal{L} = \{1, 2, \ldots, L\}$) represents the time-homogeneous Markov jump with right continuous trajectories. The transition rate matrix of $r(t)$ is described as $\Pi = [\lambda_{kl}]$ with

S. Dong et al., *Control and Filtering of Fuzzy Systems with Switched Parameters*, Studies in Systems, Decision and Control 268,
https://doi.org/10.1007/978-3-030-35566-1_5

$$\Pr\{r(t+\Delta t)=l|r(t)=k\}=\begin{cases}\lambda_{kl}\Delta t+o(\Delta t), & k\neq l,\\ 1+\lambda_{ll}\Delta t+o(\Delta t), & k=l,\end{cases} \tag{5.2}$$

where Δt is the infinitesimal transition time interval, satisfying $\lim_{\Delta t\to 0}\frac{o(\Delta t)}{\Delta t}=0$, and λ_{kl} denotes the jump rate from mode k at time t to mode l at time $t+\Delta t$ with $\lambda_{kl}\geq 0,\ k\neq l$ and $\lambda_{ll}=-\sum_{k=1,k\neq l}^{L}\lambda_{kl}$.

Via the T–S fuzzy approach, we obtain that the normalized fuzzy weighting function is $h_i(\zeta(t))=\frac{\prod_{j=1}^{p}\xi_{ij}(\zeta_j(t))}{\sum_{i=1}^{r}\prod_{j=1}^{p}\xi_{ij}(\zeta_j(t))}$ and $\zeta(t)=[\zeta_1(t),\zeta_1(t),\ldots,\zeta_p(t)]$, where $\xi_{ij}(\zeta_j(t))$ is the grade of membership $\zeta_j(t)$ in ξ_{ij}. We assume that $\prod_{j=1}^{p}\xi_{ij}(\zeta_j(t))\geq 0$, and we can easily obtain that $\sum_{i=1}^{r}h_i(\zeta(t))=1$ and $h_i(\zeta(t))\geq 0$.

To analyze conveniently, we describe $h_i(\zeta(t))$ as h_i in the following.

Accordingly, when $r(t)=k$, (5.1) is deduced as

$$\begin{cases}\dot{x}(t)=A_{kh}x(t)+B_{1kh}u(t)+D_{1kh}w(t),\\ z(t)=C_{kh}x(t)+B_{2kh}u(t)+D_{2kh}w(t),\end{cases} \tag{5.3}$$

where

$$A_{kh}=\sum_{i=1}^{r}h_iA_{ki},\ B_{1kh}=\sum_{i=1}^{r}h_iB_{1ki},$$

$$B_{2kh}=\sum_{i=1}^{r}h_iB_{2ki},\ C_{kh}=\sum_{i=1}^{r}h_iC_{ki},$$

$$D_{1kh}=\sum_{i=1}^{r}h_iD_{1ki},\ D_{2kh}=\sum_{i=1}^{r}h_iD_{2ki}.$$

The non-synchronous controller is designed by applying the PDC approach, as follows:

Controller rule i: IF $\zeta_1(t)$ is ξ_{i1}, $\zeta_2(t)$ is ξ_{i2}, ..., and $\zeta_p(t)$ is ξ_{ip}, THEN

$$u(t)=K_{\psi(t)i}x(t), \tag{5.4}$$

where $K_{\psi(t)i}\in R^{m\times n}$ is the ith fuzzy controller parameter to be solved. $\psi(t)$ is introduced to describe the HMM, taking values in $\mathcal{O}=\{1,2,\ldots,O\}$ and satisfying the conditional probability matrix $\Psi=[\phi_{ks}]$ with

$$\Pr\{\psi(t)=s|r(t)=k\}=\phi_{ks}, \tag{5.5}$$

where $\sum_{s=1}^{O}\phi_{ks}=1$.

When $\psi(t)=s$, we have

$$u(t)=K_{sh}x(t) \tag{5.6}$$

with $K_{sh}=\sum_{i=1}^{r}h_iK_{si}$.

Recalling (5.3) and (5.6), we obtain the closed-loop system:

$$\begin{cases} \dot{x}(t) = \breve{A}_{ksh}x(t) + D_{1kh}w(t), \\ z(t) = \breve{C}_{ksh}x(t) + D_{2kh}w(t), \end{cases} \tag{5.7}$$

where

$$\breve{A}_{ksh} = A_{kh} + B_{1kh}K_{sh} = \sum_{i=1}^{r}\sum_{j=1}^{r} h_i h_j \breve{A}_{ksij},$$

$$\breve{C}_{ksh} = C_{kh} + B_{2kh}K_{sh} = \sum_{i=1}^{r}\sum_{j=1}^{r} h_i h_j \breve{C}_{ksij},$$

$$\breve{A}_{ksij} = A_{ki} + B_{1ki}K_{sj}, \ \breve{C}_{ksij} = C_{ki} + B_{2ki}K_{sj},$$

$$D_{1kh} = \sum_{i=1}^{r} h_i D_{1ki}, \ D_{2kh} = \sum_{i=1}^{r} h_i D_{2ki}.$$

Based on the dissipativity theory, the energy supply function for the system (5.7) is described as

$$J(z(t), w(t), \mathcal{T}) = \int_0^{\mathcal{T}} E\{r(z(t), w(t))\}\mathrm{d}t, \tag{5.8}$$

where $r(z(t), w(t))$ is the supply rate with

$$r(z(t), w(t)) = z^T(t)Qz(t) + 2z^T(t)Sw(t) + w^T(t)Rw(t).$$

Here real matrices Q, S, and R are known with $R = R^T$ and $Q = Q^T < 0$. It can be concluded that there exists Q_- satisfying $-Q = Q_-^T Q_-$.

Definition 5.1 If for a given scalar $\alpha > 0$, any $\mathcal{T} > 0$, under the zero initial condition, the following inequality holds:

$$J(z(t), w(t), \mathcal{T}) > \alpha \int_0^{\mathcal{T}} w^T(t)w(t)\mathrm{d}t, \tag{5.9}$$

system (5.7) is said to be strictly $(Q,\ S,\ R)$-α-dissipative and α is the dissipative performance bound.

Lemma 5.1 ([1, 2]) *There exists matrix $P = P^T > 0$ such that*

$$\begin{bmatrix} PA^T + AP + X & * & * \\ B^T + DP & W_1 & * \\ CP & W_2 & W_3 \end{bmatrix} < 0 \tag{5.10}$$

if and only if there exist a scalar $\mu > 0$, matrices $P = P^T > 0$ and Z such that

$$\begin{bmatrix} -Z - Z^T & * & * & * & * \\ AZ + P & -\mu^{-1}P + X & * & * & * \\ DZ & B^T & W_1 & * & * \\ CZ & 0 & W_2 & W_3 & * \\ Z & 0 & 0 & 0 & -\mu P \end{bmatrix} < 0, \tag{5.11}$$

where real matrices $A,\ B,\ C,\ X,\ W_1,\ W_2$ *and* W_3 *are known.*

Our goal is to devise a strictly dissipative and asynchronous controller, which meets the following two requirements:
(1) System (5.7) is stochastically stable when $w(t) \equiv 0$:

$$E\left\{ \int_0^{+\infty} ||x(t)||^2 \mathrm{d}t \right\} < \infty. \tag{5.12}$$

(2) Under the zero initial condition, system (5.7) is strictly dissipative.

5.3 Main Results of Continuous-Time Systems

In this section, the stochastic stability and strictly dissipative performance of system (5.7) are considered firstly before setting out to design the controller (5.4).

Theorem 5.2 *For a prescribed* $\alpha > 0$*, if there exist matrices* $P_k > 0$ *for any* $k \in \mathcal{L}$ *and* $i, j \in \mathcal{R}$ *satisfying*

$$\Theta_{kii} < 0, \tag{5.13}$$

$$\Theta_{kij} + \Theta_{kji} < 0,\ i < j, \tag{5.14}$$

where

$$\Theta_{kij} = \begin{bmatrix} \theta_{kij}^{11} & * & * \\ \theta_{kij}^{21} & \theta_{ki}^{22} & * \\ \theta_{kij}^{31} & \theta_{ki}^{32} & -I \end{bmatrix},$$

$$\theta_{kij}^{11} = He\left(\sum_{s=1}^{O} \phi_{ks} \breve{A}_{ksij}^T P_k \right) + \sum_{l=1}^{L} \lambda_{kl} P_l,$$

$$\theta_{kij}^{21} = D_{1ki}^T P_k - S^T \left(\sum_{s=1}^{O} \phi_{ks} \breve{C}_{ksij} \right),$$

$$\theta_{ki}^{22} = -R + \alpha I - He(D_{2ki}^T S),$$

$$\theta_{kij}^{31} = \begin{bmatrix} \sqrt{\phi_{k1}} Q_- \breve{C}_{k1ij} \\ \vdots \\ \sqrt{\phi_{kO}} Q_- \breve{C}_{kOij} \end{bmatrix},\ \theta_{ki}^{32} = \begin{bmatrix} \sqrt{\phi_{k1}} Q_- D_{2ki} \\ \vdots \\ \sqrt{\phi_{kO}} Q_- D_{2ki} \end{bmatrix},$$

$$He(X) = X + X^T,$$

system (5.7) is stochastically stable with strict dissipativity.

Proof According to (5.13) and (5.14) with (5.7), it follows that

$$\begin{aligned}\Theta_{kh} &= \sum_{i=1}^{r}\sum_{j=1}^{r} h_i h_j \Theta_{kij} \\ &= \sum_{i=1}^{r} h_i^2 \Theta_{kii} + \sum_{i=1}^{r-1}\sum_{j=i+1}^{r} h_i h_j (\Theta_{kij} + \Theta_{kji}) < 0,\end{aligned} \tag{5.15}$$

where

$$\Theta_{kh} = \begin{bmatrix} \theta_{kh}^{11} & * & * \\ \theta_{kh}^{21} & \theta_{kh}^{22} & * \\ \theta_{kh}^{31} & \theta_{kh}^{32} & -I \end{bmatrix},$$

$$\theta_{kh}^{11} = \text{He}\left(\sum_{s=1}^{O} \phi_{ks} \breve{A}_{ksh}^T P_k\right) + \sum_{l=1}^{L} \lambda_{kl} P_l,$$

$$\theta_{kh}^{21} = D_{1kh}^T P_k - S^T \left(\sum_{s=1}^{O} \phi_{ks} \breve{C}_{ksh}\right),$$

$$\theta_{kh}^{22} = -R + \alpha I - \text{He}\Big(D_{2kh}^T S\Big),$$

$$\theta_{kh}^{31} = \begin{bmatrix} \sqrt{\phi_{k1}} Q_- \breve{C}_{k1h} \\ \vdots \\ \sqrt{\phi_{kO}} Q_- \breve{C}_{kOh} \end{bmatrix}, \ \theta_{kh}^{32} = \begin{bmatrix} \sqrt{\phi_{k1}} Q_- D_{2kh} \\ \vdots \\ \sqrt{\phi_{kO}} Q_- D_{2kh} \end{bmatrix}.$$

The candidate Lyapunov function is chosen as

$$V(t) = x^T(t) P_{r(t)} x(t). \tag{5.16}$$

Let $\mathcal{A}$ be the weak infinitesimal generator of the random process $\{x(t), r(t)\}$. Then we obtain

$$\mathcal{A}V(t) = x^T(t)\left(\sum_{l=1}^{L} \lambda_{kl} P_l\right) x(t) + 2x^T(t) P_k \dot{x}(t). \tag{5.17}$$

When $w(t) = 0$, considering (5.5) and (5.7), we have

$$E\{\mathcal{A}V(t)\} = x^T(t)\theta_{kh}^{11} x(t). \tag{5.18}$$

Due to (5.15), $\theta_{kh}^{11} < 0$ holds and

$$E\{\mathcal{A}V(t)\} < -x^T(t)\mu_{min}(-\theta_{kh}^{11})x(t) = -\mu||x(t)||^2, \tag{5.19}$$

where $\mu_{min}(\cdot)$ is the minimal eigenvalue of a matrix, and $\mu = \mu_{min}(-\theta_{kh}^{11})$.

Integrating (5.19) from $t = 0$ to ∞, we have

$$E\left\{\int_0^{+\infty} ||x(t)||^2 dt\right\} < \mu^{-1}E\{V(0) - V(+\infty)\} < \mu^{-1}V(0). \tag{5.20}$$

We can conclude that system (5.7) is stochastically stable.

On the other hand,

$$\begin{aligned} &E\{\mathcal{A}V(t) - r(z(t), w(t)) + \alpha w^T(t)w(t)\} \\ &= \eta^T(t)\begin{bmatrix} \theta_{kh}^{11} + (\theta_{kh}^{31})^T\theta_{kh}^{31} & * \\ \theta_{kh}^{21} + (\theta_{kh}^{32})^T\theta_{kh}^{31} & \theta_{kh}^{22} + (\theta_{kh}^{32})^T\theta_{kh}^{32} \end{bmatrix}\eta(t), \end{aligned} \tag{5.21}$$

where

$$\eta(t) = \left[x^T(t)\ w^T(t)\right]^T.$$

Applying Schur Complement to (5.15), we find that

$$\begin{bmatrix} \theta_{kh}^{11} + (\theta_{kh}^{31})^T\theta_{kh}^{31} & * \\ \theta_{kh}^{21} + (\theta_{kh}^{32})^T\theta_{kh}^{31} & \theta_{kh}^{22} + (\theta_{kh}^{32})^T\theta_{kh}^{32} \end{bmatrix} < 0. \tag{5.22}$$

Accordingly,

$$E\left\{\mathcal{A}V(t) - r(z(t), w(t)) + \alpha w^T(t)w(t)\right\} < 0. \tag{5.23}$$

By integral transformation under the zero initial condition, we have

$$E\{V(\mathcal{T})\} - \int_0^{\mathcal{T}} E\{r(z(t), w(t))\}dt + \alpha\int_0^{\mathcal{T}} w^T(t)w(t)dt < 0. \tag{5.24}$$

Owing to $V(\mathcal{T}) > 0$ and considering (5.8), $J > \alpha\int_0^{\mathcal{T}} w^T(t)w(t)dt$ holds. Based on Definition 5.1, system (5.7) is strictly dissipative. The proof is completed.

Based on the sufficient condition of Theorem 5.2 and Lemma 5.1, we find a solution to controller (5.4) for fuzzy MJSs.

Theorem 5.3 *For a prescribed $\alpha > 0$ and some scalar $\mu > 0$, if there exist matrices $G_k > 0$, Z, $\mathcal{K}_{si}$ for any $k \in \mathcal{L}$, $s \in \mathcal{O}$ and $i, j \in \mathcal{R}$ satisfying*

$$\tilde{\Theta}_{kii} < 0, \tag{5.25}$$

$$\tilde{\Theta}_{kij} + \tilde{\Theta}_{kji} < 0,\ i < j, \tag{5.26}$$

where

$$\tilde{\Theta}_{kij} = \begin{bmatrix} -Z - Z^T & * & * & * & * & * \\ \tilde{\theta}^{21}_{kij} & \tilde{\theta}^{22}_{k} & * & * & * & * \\ \tilde{\theta}^{31}_{kij} & D^T_{1ki} & \tilde{\theta}^{33}_{ki} & * & * & * \\ \tilde{\theta}^{41}_{kij} & 0 & \tilde{\theta}^{43}_{ki} & -I & * & * \\ Z & 0 & 0 & 0 & -\mu G_k & * \\ 0 & \tilde{\theta}^{62}_{k} & 0 & 0 & 0 & \tilde{\theta}^{66}_{k} \end{bmatrix},$$

$$\tilde{\theta}^{21}_{kij} = A_{ki}Z + B_{1ki}\sum_{s=1}^{O}\phi_{ks}\mathcal{K}_{sj} + G_k,$$

$$\tilde{\theta}^{22}_{k} = \lambda_{kk}G_k - \mu^{-1}G_k,$$

$$\tilde{\theta}^{31}_{kij} = -S^T C_{ki}Z - S^T B_{2ki}\sum_{s=1}^{O}\phi_{ks}\mathcal{K}_{sj},$$

$$\tilde{\theta}^{41}_{kij} = \begin{bmatrix} \sqrt{\phi_{k1}}Q_-(C_{ki}Z + B_{2ki}\mathcal{K}_{1j}) \\ \vdots \\ \sqrt{\phi_{kO}}Q_-(C_{ki}Z + B_{2ki}\mathcal{K}_{Oj}) \end{bmatrix},$$

$$\tilde{\theta}^{33}_{ki} = -R + \alpha I - He(D^T_{2ki}S),$$

$$\tilde{\theta}^{43}_{ki} = \begin{bmatrix} \sqrt{\phi_{k1}}Q_- D_{2ki} \\ \vdots \\ \sqrt{\phi_{kO}}Q_- D_{2ki} \end{bmatrix},$$

$$\tilde{\theta}^{T62}_{k} = G_k\left[\sqrt{\lambda_{k1}} \cdots \sqrt{\lambda_{kk-1}}\,\sqrt{\lambda_{kk+1}} \cdots \sqrt{\lambda_{kL}}\right],$$

$$\tilde{\theta}^{66}_{k} = -diag\{G_1, \ \ldots, G_{k-1}, G_{k+1}, \ldots, \ G_L\},$$

system (5.7) is stochastically stable with strict dissipativity. Furthermore, the controller is achieved by

$$K_{si} = \mathcal{K}_{si}Z^{-1}. \tag{5.27}$$

Proof Define $G_k = P_k^{-1} (k \in \mathcal{L})$. Due to $\tilde{\theta}^{66} < 0$, via Schur Complement, (5.25) is deduced as

$$\tilde{\Theta}'_{kii} < 0, \tag{5.28}$$

where

$$\tilde{\Theta}'_{kii} = \begin{bmatrix} -Z - Z^T & * & * & * & * \\ \tilde{\theta}^{21}_{kij} & \tilde{\theta}'^{22}_{k} & * & * & * \\ \tilde{\theta}^{31}_{kij} & D^T_{1ki} & \tilde{\theta}^{33}_{ki} & * & * \\ \tilde{\theta}^{41}_{kij} & 0 & \tilde{\theta}^{43}_{ki} & -I & * \\ Z & 0 & 0 & 0 & -\mu G_k \end{bmatrix},$$

$$\tilde{\theta}'^{22}_{k} = G_k \sum_{l=1}^{L} \lambda_{kl} G_l^{-1} G_k^T - \mu^{-1} G_k.$$

With consideration of $\mathcal{K}_{sj} = K_{sj} Z$ and Lemma 5.1, we have

$$\tilde{\Theta}''_{kii} < 0, \tag{5.29}$$

where

$$\tilde{\Theta}''_{kij} = \begin{bmatrix} \theta^{11}_{kij} & * & * \\ \tilde{\theta}'^{21}_{kij} & \theta^{22}_{ki} & * \\ \tilde{\theta}'^{31}_{kij} & \theta^{32}_{ki} & -I \end{bmatrix},$$

$$\theta^{11}_{kij} = \mathrm{He}\left(\sum_{s=1}^{O} \phi_{ks} \breve{A}_{ksij} G_k\right) + G_k \sum_{l=1}^{L} \lambda_{kl} G_l^{-1} G_k^T,$$

$$\tilde{\theta}'^{31}_{kij} = \begin{bmatrix} \sqrt{\phi_{k1}} Q_- \breve{C}_{k1ij} G_k \\ \vdots \\ \sqrt{\phi_{kO}} Q_- \breve{C}_{kOij} G_k \end{bmatrix}, \ \theta^{32}_{ki} = \begin{bmatrix} \sqrt{\phi_{k1}} Q_- D_{2ki} \\ \vdots \\ \sqrt{\phi_{kO}} Q_- D_{2ki} \end{bmatrix},$$

$$\tilde{\theta}'^{21}_{kij} = D^T_{1ki} - S^T \left(\sum_{s=1}^{O} \phi_{ks} \breve{C}_{ksij}\right) G_k,$$

$$\theta^{22}_{ki} = -R + \alpha I - \mathrm{He}(D^T_{2ki} S), \ i = j.$$

Pre- and post-multiplying $\mathrm{diag}\{G_k^{-1}, I, I\}$ with $G_k = P_k^{-1}$, (5.13) is obtained. By the same way to (5.26), (5.14) is guaranteed. The proof is completed.

Remark 5.4 With Lemma 5.1, matrices G_k and Z, we separate successfully P_k from $\breve{A}_{kij}$ that contains the controller matrix. Hence, the dissipative controller design problem can be converted into the following optimal issue in the form of LMIs. For some scalar $\mu > 0$,

$$\begin{aligned} &\min \quad -\alpha \\ &\text{s.t.} \quad \text{LMIs (5.25) and (5.26)}. \end{aligned}$$

Moreover, it is worth noting that there are $0.5r(r+1)L$ LMIs to be solved and $(0.5L+1)n^2 + (0.5L + rmO)n$ variables to be determined.

5.4 Preliminary Analysis of Discrete-Time Systems

Consider the following nonlinear MJSs, described via the T–S fuzzy approach.
Plant rule i: IF ϑ_{1k} is η_{i1}, ϑ_{2k} is η_{i2}, ..., and ϑ_{lk} is η_{il}, THEN

$$\begin{cases} x_{k+1} = A_{r_k i} x_k + B_{1 r_k i} u_k + D_{1 r_k i} w_k, \\ z_k = C_{r_k i} x_k + B_{2 r_k i} u_k + D_{2 r_k i} w_k, \end{cases} \tag{5.30}$$

where $i \in \mathcal{I} = \{1, 2, \ldots, r\}$ is the ith fuzzy rule and r is the total number of rules; ϑ_{jk} $(j \in \{1, 2, \ldots, l\})$ is the premise variable; η_{ij} is the fuzzy set. $x_k \in R^n$ is the state variable; $u_k \in R^m$ is the control input; $z_k \in R^p$ is the controlled output and $w_k \in R^o$ is the external disturbance which belongs to $l_2[0, +\infty)$. $A_{r_k i}$, $B_{1 r_k i}$, $B_{2 r_k i}$, $C_{r_k i}$, $D_{1 r_k i}$, and $D_{2 r_k i}$ are given with appropriate dimensions. $r_k \in \mathcal{M} = \{1, 2, \ldots, M\}$ is applied to represent the Markov jump phenomenon. The transition probability matrix of jumps is denoted as $\Pi = [\pi_{st}]$ and the jump r_k is subject to

$$\Pr\{r_{k+1} = t | r_k = s\} = \pi_{st}, \; s, \; t \in \mathcal{M}, \tag{5.31}$$

It is easy to find that $0 \leq \pi_{st} \leq 1$ and $\sum_{t=1}^{M} \pi_{st} = 1$.

Through the T–S fuzzy inference, the overall considered system with $r_k = s$ is inferred as

$$\begin{cases} x_{k+1} = A_{sh} x_k + B_{1sh} u_k + D_{1sh} w_k, \\ z_k = C_{sh} x_k + B_{2sh} u_k + D_{2sh} w_k, \end{cases} \tag{5.32}$$

where

$$\begin{aligned}
A_{sh} &= \sum_{i=1}^{r} h_i(\vartheta_k) A_{si}, \; C_{sh} = \sum_{i=1}^{r} h_i(\vartheta_k) C_{si}, \\
B_{1sh} &= \sum_{i=1}^{r} h_i(\vartheta_k) B_{1si}, \; B_{2sh} = \sum_{i=1}^{r} h_i(\vartheta_k) B_{2si}, \\
D_{1sh} &= \sum_{i=1}^{r} h_i(\vartheta_k) D_{1si}, \; D_{2sh} = \sum_{i=1}^{r} h_i(\vartheta_k) D_{2si}, \\
\vartheta_k &= \left[\vartheta_{1k}, \vartheta_{2k}, \ldots, \vartheta_{lk}\right], \; h_i(\vartheta_k) = \frac{\prod_{j=1}^{l} \eta_{ij}(\vartheta_{jk})}{\sum_{i=1}^{r} \prod_{j=1}^{l} \eta_{ij}(\vartheta_{jk})}.
\end{aligned}$$

The variable $\eta_{ij}(\vartheta_{jk})$ represents the grade of membership ϑ_{jk} in η_{ij}. $h_i(\vartheta_k)$ is the normalized fuzzy weighting function. Here, we suppose that $\prod_{j=1}^{l} \eta_{ij}(\vartheta_{jk}) \geq 0$. It easily follows that

$$h_i(\vartheta_k) \geq 0, \; \sum_{i=1}^{r} h_i(\vartheta_k) = 1. \tag{5.33}$$

For notational brevity, $h_i(\vartheta_k)$ is described as h_i in the later section.

In this chapter, by the PDC approach, the fuzzy asynchronous state-feedback controller is devised as follows.

Controller rule i: IF ϑ_{1k} is η_{i1}, ϑ_{2k} is η_{i2}, ..., and ϑ_{lk} is η_{il}, THEN

$$u(k) = K_{\delta_k i} x(k), \tag{5.34}$$

where $K_{\delta_k i} \in R^{m\times n}$ is the ith local state-feedback controller parameter to be determined. δ_k has the same effect as r_k, taking values in $\mathcal{N} = \{1, 2, \ldots, N\}$ and satisfying the conditional probability matrix $\Omega = [\varphi_{sv}]$ with

$$\Pr\{\delta_k = v | r_k = s\} = \varphi_{sv}, \ s \in \mathcal{M}, \ v \in \mathcal{N}, \tag{5.35}$$

where $\sum_{v=1}^{N} \varphi_{sv} = 1$ and $0 \le \varphi_{sv} \le 1$.

Then, under $\delta_k = v$, it follows that

$$u(k) = K_{vh} x(k), \tag{5.36}$$

where

$$K_{vh} = \sum_{i=1}^{r} h_i K_{vi}.$$

The following closed-loop fuzzy system is achieved by combining system (5.32) and asynchronous controller (5.36):

$$\begin{cases} x_{k+1} = \bar{A}_{svh} x_k + D_{1sh} w_k, \\ z_k = \bar{C}_{svh} x_k + D_{2sh} w_k, \end{cases} \tag{5.37}$$

where

$$\begin{aligned}
\bar{A}_{svh} &= A_{sh} + B_{1sh} K_{vh} = \sum_{i=1}^{r}\sum_{j=1}^{r} h_i h_j \bar{A}_{svij}, \\
\bar{C}_{svh} &= C_{sh} + B_{2sh} K_{vh} = \sum_{i=1}^{r}\sum_{j=1}^{r} h_i h_j \bar{C}_{svij}, \\
\bar{A}_{svij} &= A_{si} + B_{1si} K_{vj}, \ \bar{C}_{svij} = C_{si} + B_{2si} K_{vj}, \\
D_{1sh} &= \sum_{i=1}^{r} h_i(\vartheta_k) D_{1si}, \ D_{2sh} = \sum_{i=1}^{r} h_i(\vartheta_k) D_{2si}.
\end{aligned}$$

The following quadratic energy supply rate for input-output pairs $(w_k, \ z_k)$ in system (5.37) is described as

$$r(z_k, w_k) = z_k^T \mathcal{Q} z_k + 2z_k^T \mathcal{S} w_k + w_k^T \mathcal{R} w_k. \tag{5.38}$$

Here $\mathcal{Q}$, $\mathcal{S}$ and $\mathcal{R}$ are given real matrices; $\mathcal{R}$ satisfies $\mathcal{R}^T = \mathcal{R}$; $\mathcal{Q}$ is negative semi-definite matrices, which implies that $-\mathcal{Q} = \mathcal{Q}_-^T \mathcal{Q}_-$.

From (5.38), we obtain the quadratic energy supply function for the system (5.37) in the following.

$$J(z_k, w_k, T) = \sum_{k=0}^{T} E\{r(z_k, w_k)\}. \tag{5.39}$$

The following definition is introduced to facilitate investigation later.

Definition 5.2 ([3]) If for any $T > 0$, under the zero initial condition, the following inequality holds:

$$J(z_k, w_k, T) > 0, \tag{5.40}$$

system (5.37) is said to be $(\mathcal{Q}, \mathcal{S}, \mathcal{R})$-dissipative regarding the supply rate (5.38). Moreover, if for a given scalar $\alpha > 0$ and any $T > 0$, under the zero initial condition, the following inequality holds:

$$J(z_k, w_k, T) > \alpha \sum_{k=0}^{T} w_k^T w_k, \tag{5.41}$$

system (5.37) is said to be strictly $(\mathcal{Q}, \mathcal{S}, \mathcal{R})$-$\alpha$-dissipative regarding the supply rate (5.38) and α is the dissipativity performance bound.

The major purpose of our chapter is to devise the controller (5.36) guaranteeing that the closed-loop system (5.37) satisfies:
(1) System (5.37) is stochastically stable when $w_k \equiv 0$, that is,

$$E\left\{\sum_{k=0}^{\infty} ||x_k||^2 | x_0, r_0, \delta_0\right\} < \infty; \tag{5.42}$$

(2) Under the zero initial condition, system (5.37) is strictly $(\mathcal{Q}, \mathcal{S}, \mathcal{R})$-$\alpha$-dissipative.

5.5 Main Results of Discrete-Time Systems

In the section, a sufficient condition is firstly presented to prove the stability and strict dissipativity of system (5.37) on the assumption that controller gain K_{vi} is known. Then, we focus on K_{vi} design for system (5.37).

Theorem 5.5 *If there exist matrices $P_{si} > 0$ and $W_{svi} > 0$ for any $s \in \mathcal{M}$, $v \in \mathcal{N}$ and $a, i, j \in \mathcal{I}$ subject to*

$$\sum_{v=1}^{N} \varphi_{sv} W_{svi} < P_{si}, \tag{5.43}$$

$$\Sigma_{svaii} < 0, \tag{5.44}$$

$$\Sigma_{svaij} + \Sigma_{svaji} < 0, \; i < j, \tag{5.45}$$

system (5.37) is stochastically stable and strictly $(\mathcal{Q}, \; \mathcal{S}, \; \mathcal{R})$*-*$\alpha$*-dissipative, where*

$$\Sigma_{svaij} = \begin{bmatrix} -W_{svi} & -\bar{C}_{svij}^T \mathcal{S} & \bar{C}_{svij}^T \mathcal{Q}_-^T & \Sigma_{14}^T \\ * & \Sigma_{22} & D_{2si}^T \mathcal{Q}_-^T & \Sigma_{24}^T \\ * & * & -I & 0 \\ * & * & * & -\Sigma_{44} \end{bmatrix},$$

$$\Sigma_{22} = -\mathcal{R} + \alpha I - D_{2si}^T \mathcal{S} - \mathcal{S}^T D_{2si},$$

$$\Sigma_{44} = diag\{P_{1a}, \; P_{2a}, \ldots, \; P_{Ma}\},$$

$$\Sigma_{14} = \begin{bmatrix} \sqrt{\pi_{s1}} P_{1a} \bar{A}_{svij}, \\ \sqrt{\pi_{s2}} P_{2a} \bar{A}_{svij} \\ \vdots \\ \sqrt{\pi_{sM}} P_{Ma} \bar{A}_{svij} \end{bmatrix}, \; \Sigma_{24} = \begin{bmatrix} \sqrt{\pi_{s1}} P_{1a} D_{1si} \\ \sqrt{\pi_{s2}} P_{2a} D_{1si} \\ \vdots \\ \sqrt{\pi_{sM}} P_{Ma} D_{1si} \end{bmatrix},$$

$$-\mathcal{Q} = \mathcal{Q}_-^T \mathcal{Q}_-.$$

Proof Firstly, define

$$\begin{aligned} P_{sh} &= \sum_{i=1}^{r} h_i P_{si}, \; P_{th^+} = \sum_{a=1}^{r} h_a^+ P_{ta}, \\ W_{svh} &= \sum_{i=1}^{r} h_i W_{svi}, \; h^+ = h_{k+1}, \; t \in \mathcal{M}, \end{aligned} \tag{5.46}$$

where h^+ denotes the normalized fuzzy weighting function at time $k+1$.

From (5.43), we have

$$\sum_{v=1}^{N} \varphi_{sv} W_{svh} < P_{sh}. \tag{5.47}$$

And then together with system (5.37), (5.44) and (5.45), it follows that

$$
\begin{aligned}
\Sigma_{svh} &= \sum_{a=1}^{r}\sum_{i=1}^{r}\sum_{j=1}^{r} h_a^+ h_i h_j \Sigma_{svaij} = \sum_{a=1}^{r} h_a^+ \left(\sum_{i=1}^{r} h_i^2 \Sigma_{svaii} \right. \\
&\left. + \sum_{i=1}^{r-1}\sum_{j=i+1}^{r} h_i h_j (\Sigma_{svaij} + \Sigma_{svaji}) \right) < 0,
\end{aligned} \tag{5.48}
$$

where

$$
\begin{aligned}
\Sigma_{svah} &= \begin{bmatrix} -W_{svh} & -\bar{C}_{svh}^T \mathcal{S} & \bar{C}_{svh}^T \mathcal{Q}_-^T & \Sigma_{14}^{'T} \\ * & \Sigma_{22}^{'} & D_{2sh}^T \mathcal{Q}_-^T & \Sigma_{24}^{'T} \\ * & * & -I & 0 \\ * & * & * & -\Sigma_{44}^{'} \end{bmatrix}, \\
\Sigma_{22}^{'} &= -\mathcal{R} + \alpha I - D_{2sh}^T \mathcal{S} - \mathcal{S}^T D_{2sh}, \\
\Sigma_{44}^{'} &= \mathrm{diag}\{P_{1h^+},\ P_{2h^+}, \ldots,\ P_{Mh^+}\}, \\
\Sigma_{14}^{'} &= \begin{bmatrix} \sqrt{\pi_{s1}} P_{1h^+} \bar{A}_{svh} \\ \sqrt{\pi_{s2}} P_{2h^+} \bar{A}_{svh} \\ \vdots \\ \sqrt{\pi_{sM}} P_{Mh^+} \bar{A}_{svh} \end{bmatrix}, \ \Sigma_{24}^{'} = \begin{bmatrix} \sqrt{\pi_{s1}} P_{1h^+} D_{1sh} \\ \sqrt{\pi_{s2}} P_{2h^+} D_{1sh} \\ \vdots \\ \sqrt{\pi_{sM}} P_{Mh^+} D_{1sh} \end{bmatrix}.
\end{aligned}
$$

Owing to $P_{th^+} = \sum_{a=1}^{r} h_a^+ P_{ta}$ and $P_{ta} > 0$, the following inequalities are obtained by Schur Complement:

$$
\bar{A}_{svh}^T X_{sh^+} \bar{A}_{svh} - W_{svh} < 0, \tag{5.49}
$$

and

$$
\psi_{1svh}^T X_{sh^+} \psi_{1svh} - \psi_{2svh}^T \mathcal{Q} \psi_{2svh} + \psi_{3svh} < 0, \tag{5.50}
$$

where

$$
\begin{aligned}
\psi_{1svh} &= \begin{bmatrix} \bar{A}_{svh} & D_{1sh} \end{bmatrix},\ \psi_{2svh} = \begin{bmatrix} \bar{C}_{svh} & D_{2sh} \end{bmatrix}, \\
\psi_{3svh} &= \begin{bmatrix} -W_{svh} & -\bar{C}_{svh}^T \mathcal{S} \\ * & \Sigma_{22}^{'} \end{bmatrix},\ X_{sh^+} = \sum_{t=1}^{M} \pi_{st} P_{th^+}.
\end{aligned}
$$

According to (5.47), (5.49) and (5.50), it is clear to find that

$$
\sum_{v=1}^{N} \varphi_{sv} \bar{A}_{svh}^T X_{sh^+} \bar{A}_{svh} - P_{sh} < 0, \tag{5.51}
$$

and

$$
\sum_{v=1}^{N} \varphi_{sv} \left(\psi_{1svh}^T X_{sh^+} \psi_{1svh} - \psi_{2svh}^T \mathcal{Q} \psi_{2svh} + \psi_{3svh}^{'} \right) < 0, \tag{5.52}
$$

where

$$\psi'_{3svh} = \begin{bmatrix} -P_{sh} & -\bar{C}^T_{svh}\mathcal{S} \\ * & \Sigma'_{22} \end{bmatrix}.$$

Construct the following Lyapunov function:

$$V_k = x_k^T P_{sh} x_k. \tag{5.53}$$

Along system (5.37) with $w_k \equiv 0$, we have

$$\begin{aligned} E\{\Delta V_k\} &= E\{x_{k+1}^T P_{th^+} x_{k+1}\} - E\{x_k^T P_{sh} x_k\} \\ &= x_k^T \left(\sum_{v=1}^{N} \varphi_{sv} \bar{A}^T_{svh} X_{sh^+} \bar{A}_{svh} - P_{sh} \right) x_k. \end{aligned} \tag{5.54}$$

Recalling (5.51), we acquire

$$\begin{aligned} E\{\Delta V_k\} &< -\lambda_{min} \left(-\sum_{v=1}^{N} \varphi_{sv} \bar{A}^T_{svh} X_{sh^+} \bar{A}_{svh} + P_{sh} \right) x_k^T x_k \\ &< -\beta x_k^T x_k \\ &< 0, \end{aligned} \tag{5.55}$$

where

$$\beta = \inf \left\{ \lambda_{min} \left(-\sum_{v=1}^{N} \varphi_{sv} \bar{A}^T_{svh} X_{sh^+} \bar{A}_{svh} + P_{sh} \right),\ s \in \mathcal{M} \right\}.$$

From $k = 0$ to ∞, summing up the inequality of both sides, we have

$$\begin{aligned} E\left\{ \sum_{k=0}^{\infty} ||x_k||^2 | x_0, r_0, \delta_0 \right\} &< \frac{1}{\beta}(E\{V_0\} - E\{V_\infty\}) \\ &< \frac{1}{\beta} E\{V_0\} \\ &< \infty. \end{aligned} \tag{5.56}$$

According to (5.42), (5.56) implies that system (5.37) is stochastically stable.

As for the supply rate (5.38), we have

$$\begin{aligned} &E\{\Delta V_k - r(z_k, w_k) + \alpha w_k^T w_k\} \\ &= \zeta_k^T \sum_{v=1}^{N} \varphi_{sv} \left(\psi^T_{1svh} X_{sh^+} \psi_{1svh} - \psi^T_{2svh} \mathcal{Q} \psi_{2svh} + \psi'_{3svh} \right) \zeta_k, \end{aligned} \tag{5.57}$$

where

$$\zeta_k = \left[x_k^T \ w_k^T\right]^T.$$

From $k = 0$ to T, summing up both side and recalling (5.39) and (5.52), we obtain that

$$V_{T+1} - V_0 - J(z_k, w_k, T) + \alpha \sum_{k=0}^{T} w_k^T w_k < 0. \tag{5.58}$$

With the zero initial condition, namely, $V_0 = 0$, it follows that

$$\begin{aligned} J(z_k, w_k, T) &> \alpha \sum_{k=0}^{T} w_k^T w_k + V_{T+1} \\ &> \alpha \sum_{k=0}^{T} w_k^T w_k. \end{aligned} \tag{5.59}$$

According to Definition 5.2, we have that system (5.37) is strictly ($\mathcal{Q}$, $\mathcal{S}$, $\mathcal{R}$)-α-dissipative. The proof is finished.

Remark 5.6 From the energy dissipativity aspect, V_k is the internally stored energy function of a system, and $J(z_k, w_k, T)$ is regarded as the supplied energy from external environment with the supply rate $r(z_k, w_k)$ from time 0 to T. If the supplied energy by $r(z_k, w_k)$ is more than the stored energy inside a system, the system is said to be dissipative [4, 5].

In the following, we investigate the control design method for the controller gain K_{vi} in (5.36) based on Theorem 5.5.

Theorem 5.7 *If there exist matrices $\tilde{P}_{si} > 0$, $\tilde{W}_{svi} > 0$, G_v and $\tilde{K}_{vi}$ for any $s \in \mathcal{M}$, $v \in \mathcal{N}$ and $a, i, j \in \mathcal{I}$ subject to*

$$\begin{bmatrix} -\tilde{P}_{si} & \tilde{\mathcal{P}}_{si} \\ * & -\tilde{\mathcal{W}}_{svi} \end{bmatrix} < 0, \tag{5.60}$$

$$\Gamma_{svaii} < 0, \tag{5.61}$$

$$\Gamma_{svaij} + \Gamma_{svaji} < 0, \ i < j, \tag{5.62}$$

system (5.37) is stochastically stable and strictly ($\mathcal{Q}$, $\mathcal{S}$, $\mathcal{R}$)-α-dissipative. Moreover, the controller gain in (5.36) can be obtained as

$$K_{vj} = \tilde{K}_{vj} G_v^{-1}, \tag{5.63}$$

where

$$\tilde{\mathcal{P}}_{si} = \tilde{P}_{si}\left[\sqrt{\varphi_{s1}}\ \sqrt{\varphi_{s2}}\ \cdots\ \sqrt{\varphi_{sN}}\right],$$

$$\tilde{\mathcal{W}}_{svi} = diag\{\tilde{W}_{s1i},\ \tilde{W}_{s2i},\ldots,\ \tilde{W}_{sNi}\},$$

$$\Gamma_{svaij} = \begin{bmatrix} \Gamma_{11} & -\Gamma_{12}^T\mathcal{S} & \Gamma_{12}^T\mathcal{Q}_-^T & \Gamma_{14}^T \\ * & \Gamma_{22} & D_{2si}^T\mathcal{Q}_-^T & \Gamma_{24}^T \\ * & * & -I & 0 \\ * & * & * & -\Gamma_{44} \end{bmatrix},$$

$$\Gamma_{11} = \tilde{W}_{svi} - G_v - G_v^T,\ \Gamma_{12} = C_{si}G_v + B_{2si}\tilde{K}_{vj},$$
$$\Gamma_{22} = -\mathcal{R} + \alpha I - D_{2si}^T\mathcal{S} - \mathcal{S}^T D_{2si},$$
$$\Gamma_{44} = diag\{\tilde{P}_{1a},\ \tilde{P}_{2a},\ldots,\ \tilde{P}_{Ma}\},$$

$$\Gamma_{14} = \begin{bmatrix} \sqrt{\pi_{s1}}(A_{si}G_v + B_{1si}\tilde{K}_{vj}) \\ \sqrt{\pi_{s2}}(A_{si}G_v + B_{1si}\tilde{K}_{vj}) \\ \vdots \\ \sqrt{\pi_{sM}}(A_{si}G_v + B_{1si}\tilde{K}_{vj}) \end{bmatrix},$$

$$\Gamma_{24} = \begin{bmatrix} \sqrt{\pi_{s1}}D_{1si} \\ \sqrt{\pi_{s2}}D_{1si} \\ \vdots \\ \sqrt{\pi_{sM}}D_{1si} \end{bmatrix},\quad -\mathcal{Q} = \mathcal{Q}_-^T\mathcal{Q}_-.$$

Proof Owing to $\tilde{W}_{svi} > 0$, the following inequality is obtained by Schur Complement:

$$\sum_{v=1}^{N}\varphi_{sv}\tilde{P}_{si}\tilde{W}_{svi}^{-1}\tilde{P}_{si}^T < \tilde{P}_{si}, \tag{5.64}$$

where

$$\tilde{W}_{svi} = W_{svi}^{-1},\ \tilde{P}_{si} = P_{si}^{-1}.$$

Via pre- and post-multiplying $\tilde{P}_{si}^{-1}$ to (5.64), we obtain (5.43), which means that (5.60) is equivalent to (5.43). Pre- and post-multiplying diag$\{I, I, I, \Gamma_{44}^{-1}\}$ to (5.61), the following inequality is obtained:

$$\Gamma_{svaii}^1 < 0, \tag{5.65}$$

where

$$
\begin{aligned}
\Gamma^1_{svaii} &= \begin{bmatrix} \Gamma_{11} & -\Gamma_{12}^T S & \Gamma_{12}^T \mathcal{Q}_-^T & \Gamma_{14}^{'T} \\ * & \Gamma_{22} & D_{2si}^T \mathcal{Q}_-^T & \Gamma_{24}^{'T} \\ * & * & -I & 0 \\ * & * & * & -\Gamma_{44}' \end{bmatrix}, \\
\Gamma_{44}' &= \mathrm{diag}\{P_{1a},\ P_{2a}, \ldots,\ P_{Ma}\}, \\
\Gamma_{14}' &= \begin{bmatrix} \sqrt{\pi_{s1}} P_{1a}(A_{si} G_v + B_{1si} \tilde{K}_{vj}) \\ \sqrt{\pi_{s2}} P_{2a}(A_{si} G_v + B_{1si} \tilde{K}_{vj}) \\ \vdots \\ \sqrt{\pi_{sM}} P_{Ma}(A_{si} G_v + B_{1si} \tilde{K}_{vj}) \end{bmatrix}, \\
\Gamma_{24}' &= \begin{bmatrix} \sqrt{\pi_{s1}} P_{1a} D_{1si} \\ \sqrt{\pi_{s2}} P_{2a} D_{1si} \\ \vdots \\ \sqrt{\pi_{sM}} P_{Ma} D_{1si} \end{bmatrix},\ i = j.
\end{aligned}
$$

Because of $\tilde{W}_{svi} > 0$, we have

$$
(\tilde{W}_{svi} - G_v^T)\tilde{W}_{svi}^{-1}(\tilde{W}_{svi} - G_v) > 0. \tag{5.66}
$$

Together with $\tilde{W}_{svi} = W_{svi}^{-1}$, we obtain that

$$
-G_v^T W_{svi} G_v < \tilde{W}_{svi} - G_v - G_v^T. \tag{5.67}
$$

It is easy to find that

$$
\Gamma^2_{svaii} < 0, \tag{5.68}
$$

where

$$
\begin{aligned}
\Gamma^2_{svaii} &= \begin{bmatrix} \Gamma_{11}' & -\Gamma_{12}^T S & \Gamma_{12}^T \mathcal{Q}_-^T & \Gamma_{14}^{'T} \\ * & \Gamma_{22} & D_{2si}^T \mathcal{Q}_-^T & \Gamma_{24}^{'T} \\ * & * & -I & 0 \\ * & * & * & -\Gamma_{44}' \end{bmatrix}, \\
\Gamma_{11}' &= -G_v^T W_{svi} G_v.
\end{aligned}
$$

Through pre-multiplying $\mathrm{diag}\{G_v^{-T}, I, I, I\}$ and post-multiplying $\mathrm{diag}\{G_v^{-1}, I, I, I\}$ to (5.68), respectively, we have (5.44) with consideration of $\tilde{K}_{vj} = K_{vj} G_v$. It means that (5.61) is sufficient for (5.44). Adopting the similar proof process, we clearly find out that (5.45) can be deduced from (5.62). Furthermore, if there is a solution to (5.60)–(5.62), the controller gain in (5.36) is inferred as $K_{vj} = \tilde{K}_{vj} G_v^{-1}$. The proof is finished.

Remark 5.8 In (5.44)–(5.45), it is quite difficult to obtain the controller gain K_{vj} because of nonlinear terms like $P_{1a}\bar{A}_{svij}$ ($\bar{A}_{svij} = A_{si} + B_{1si} K_{vj}$), where P_{1a} and K_{vj} need determining. To overcome this issue, we introduce a mode-dependent slack matrix G_v and apply the relaxation approach in (5.67) to simplify a complicated nonlinear control problem into the feasibility of a set of LMIs, which can be solved by LMI Toolbox in Matlab.

Remark 5.9 Theorem 5.7 presents a sufficient condition to guarantee that system (5.37) is stochastically stable with strictly $(\mathcal{Q}, \mathcal{S}, \mathcal{R})$-$\alpha$-dissipative performance criterion in the form of LMIs. The optimal dissipative performance α^* and $(\mathcal{Q}, \mathcal{S}, \mathcal{R})$-$\alpha^*$-dissipative controller (5.36) can be achieved via solving a convex optimization problem, as shown below:

$$\begin{aligned} &\min \quad -\alpha \\ &\text{s.t.} \quad (5.60),\ (5.61), \text{and } (6.62). \end{aligned}$$

Remark 5.10 It is necessary to note that the achieved results can be extended into two special cases:
(1) H_∞ performance by letting $\mathcal{Q} = -I,\ \mathcal{S} = 0$, and $\mathcal{R} = (\alpha^2 + \alpha)I$ in (5.61) and (5.62);
(2) Passivity performance by letting $\mathcal{Q} = 0,\ \mathcal{S} = I$, and $\mathcal{R} = 2\alpha I$ in (5.61) and (5.62) if z_k has the same dimension as w_k.

In addition, in Theorem 5.7, if $\mathcal{I} = \{1\}, \mathcal{Q} = 0,\ \mathcal{S} = I$ and $\mathcal{R} = 2\alpha I$, the passivity issue has been studied in [6].

5.6 Illustrative Example

Example 1 To verify the correctness of our approach, we provide a mass-spring-damper mechanical system in [1, 7, 8], as follows:

$$M\ddot{y}(t) + g(y(t), \dot{y}(t)) + f(y(t)) = \phi(\dot{y}(t))u(t) + w(t),$$

where M is the mass; $y(t)$ is the position; $u(t)$ is the force and $w(t)$ is the disturbance. Nonlinear terms $g(y(t), \dot{y}(t))$, $f(y(t))$ and $\phi(\dot{y}(t))$ are respectively related with the damper, the spring and the input. Assume that $M = 1.0$, $g(y(t), \dot{y}(t)) = \dot{y}(t)$, $f(y(t)) = c(t)y(t)$ $(c(t) \in [0.5,\ 1.81])$, and $\phi(\dot{y}(t)) = 1 + 0.13\dot{y}^3(t)$. Here we choose $x(t) = \begin{bmatrix}\dot{y}^T(t) & y^T(t)\end{bmatrix}^T$.

Due to parameter or structure variations in the system, we assume that there are 2 jump modes occurring with the following transition rate matrix:

$$\Pi = \begin{bmatrix} -3 & 3 \\ 4 & -4 \end{bmatrix}.$$

Through T–S fuzzy inference, the original system is inferred as system (5.1) with the following parameters:

$$A_{11} = \begin{bmatrix} -1.0 & -1.555 \\ 1 & 0 \end{bmatrix}, \; B_{111} = \begin{bmatrix} 1.4387 \\ 0 \end{bmatrix},$$
$$A_{12} = \begin{bmatrix} -1.0 & -1.155 \\ 1 & 0 \end{bmatrix}, \; B_{112} = \begin{bmatrix} 0.5613 \\ 0 \end{bmatrix},$$
$$A_{21} = \begin{bmatrix} -1.0 & -2.210 \\ 1 & 0 \end{bmatrix}, \; B_{121} = \begin{bmatrix} 0.5755 \\ 0 \end{bmatrix},$$
$$A_{22} = \begin{bmatrix} -1.0 & -2.210 \\ 1 & 0 \end{bmatrix}, \; B_{122} = \begin{bmatrix} 0.2245 \\ 0 \end{bmatrix},$$
$$B_{211} = B_{212} = \begin{bmatrix} 0 \\ 1 \end{bmatrix}, \; B_{221} = B_{222} = \begin{bmatrix} 0 \\ 0.8 \end{bmatrix},$$
$$C_{11} = C_{12} = C_{21} = C_{22} = \begin{bmatrix} 0 & 1 \\ 0 & 0 \end{bmatrix},$$
$$D_{111} = D_{112} = D_{121} = D_{122} = \begin{bmatrix} 1 \\ 0 \end{bmatrix},$$
$$D_{211} = D_{212} = D_{221} = D_{222} = \begin{bmatrix} 0 \\ 0 \end{bmatrix},$$
$$h_1 = 0.5 + \frac{x_1^3(t)}{6.75}, \; h_2 = 1 - h_1.$$

The conditional probability matrix Ψ, and dissipative parameters are

$$\Psi = \begin{bmatrix} 0.2 & 0.8 \\ 0.4 & 0.6 \end{bmatrix}, \; Q = \begin{bmatrix} -0.49 & 0 \\ 0 & -0.49 \end{bmatrix},$$
$$S = \begin{bmatrix} 0.1 \\ 0.1 \end{bmatrix}, \; R = 1.$$

By solving LMI-based inequalities in Theorem 5.3 with $\mu = 1$, the optimal dissipativity index is $\alpha^* = 0.6011$, and the corresponding controller matrices are

$$K_{11} = \begin{bmatrix} -0.2178 & -0.1975 \end{bmatrix}, \; K_{12} = \begin{bmatrix} -0.1152 & -0.1914 \end{bmatrix},$$
$$K_{21} = \begin{bmatrix} -0.1130 & 0.0819 \end{bmatrix}, \; K_{22} = \begin{bmatrix} -0.1110 & -0.2061 \end{bmatrix}.$$

Under the external disturbance $w(t) = e^{-0.2t}$, Fig. 5.1 shows that $x(t)$ approaches to zero over time with the initial state $x(0) = \begin{bmatrix} -1 & -0.5 \end{bmatrix}^T$.

Table 5.3 shows that three optimal indexes α^* change with different non-synchronous degrees. Tables 5.1 and 5.2 present the corresponding conditional probability matrices Ψ and dissipative parameters (Q, S, R), respectively. From Table 5.3, we can easily find out that with asynchronous intensification, the optimal α^* of dissipativity index becomes smaller. There is a different trend to H_∞ and passivity indexes: α^* of them becomes bigger with an increasing asynchronous level.

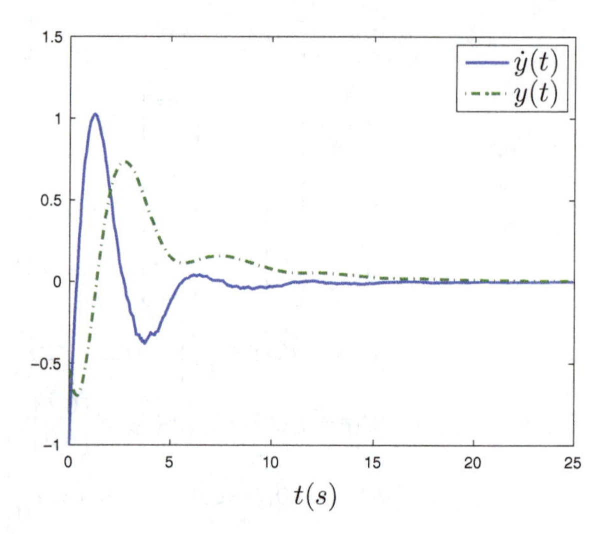

Fig. 5.1 State trajectories for the system (5.7)

Table 5.1 Conditional probability matrices Ψ

	Case I	Case II	Case III
Ψ	$\begin{bmatrix} 1 & 0 \\ 0 & 1 \end{bmatrix}$	$\begin{bmatrix} 1 & 0 \\ 0.4 & 0.6 \end{bmatrix}$	$\begin{bmatrix} 0.2 & 0.8 \\ 0.4 & 0.6 \end{bmatrix}$

Table 5.2 Q, S and R for various criteria

	Q	S	R
Dissapativity	$\begin{bmatrix} -1 & 0 \\ 0 & -1 \end{bmatrix}$	$\begin{bmatrix} 0.1 \\ 0.1 \end{bmatrix}$	1
H_∞	$\begin{bmatrix} -1 & 0 \\ 0 & -1 \end{bmatrix}$	0	$\alpha^2 + \alpha$
Passivity	0	$\begin{bmatrix} 1 \\ 1 \end{bmatrix}$	2α

Table 5.3 Optimal α^* for various criteria with different conditional probabilities

α^*	Case I	Case II	Case III
Dissapativity	0.1933	0.1880	0.1741
H_∞	0.9046	0.9074	0.9159
Passivity(10^{-9})	1.6687	1.6899	1.7099

Remark 5.11 If the output $z(t)$ has the same dimension as the external noise $w(t)$, we can study the passive problem by adjusting $Q = 0,\ S = I$, and $R = 2\alpha$. In the simulation example, it should be noted that we set $S = [1,\ 1]^T$ to adjust the dimension incompatibility problem between $z(t)$ and $w(t)$ for investigating the asynchronous passive character.

Example 2 The single-link robot arm system in Chap. 4 is used to verify the correctness of the developed design approach for discrete-time fuzzy systems. In this example, the external noise is considered. The system parameters are given as follows:

$$
\begin{aligned}
A_{r_k 1} &= \begin{bmatrix} 1 & T \\ -\frac{TglM_{r_k}}{J_{r_k}} & 1-\frac{TR}{J_{r_k}} \end{bmatrix},\ A_{r_k 2} = \begin{bmatrix} 1 & T \\ -\frac{\beta TglM_{r_k}}{J_{r_k}} & 1-\frac{TR}{J_{r_k}} \end{bmatrix}, \\
B_{1r_k 1} &= B_{1r_k 2} = \begin{bmatrix} 0 \\ \frac{T}{J_{r_k}} \end{bmatrix},\ D_{1r_k 1} = D_{1r_k 2} = \begin{bmatrix} 0 \\ T \end{bmatrix}, \\
B_{2r_k 1} &= B_{2r_k 2} = 0,\ C_{r_k 1} = C_{r_k 2} = \begin{bmatrix} 1 & 0 \end{bmatrix}, \\
D_{2r_k 1} &= D_{2r_k 2} = 0.1,\ r_k = \{1, 2, 3, 4\}, \\
g &= 9.81,\ R = 2,\ l = 0.5,\ \beta = 10^{-2}/\pi, \\
h_1(x_1(k)) &= \begin{cases} \dfrac{\sin(x_1(k)) - \beta x_1(k)}{(1-\beta)x_1(k)}, & x_1(k) \neq 0, \\ 1, & x_1(k) = 0, \end{cases} \\
h_2(x_1(k)) &= 1 - h_2(x_1(k)),
\end{aligned}
$$

where $x_1(k) \in (-\pi,\ \pi)$ and $\beta = 10^{-2}/\pi$. M_{r_t} and J_{r_t} are the payload mass and the inertia moment, varying with Markov jump r_t. The transition probability matrix is assumed to be

$$
\Pi = \begin{bmatrix} 0.3 & 0.2 & 0.4 & 0.1 \\ 0.4 & 0.2 & 0.2 & 0.2 \\ 0.55 & 0.15 & 0.3 & 0 \\ 0.1 & 0.2 & 0.3 & 0.4 \end{bmatrix}.
$$

Values M_{r_t} and J_{r_t} are shown in Table 5.4.

We assume that the conditional probability matrix of Markov jump δ_k is

Table 5.4 Values M_{r_t} and J_{r_t} for different modes r_t

r_t	1	2	3	4
M_{r_t}	1	5	10	15
J_{r_t}	1	5	10	15

$$\Omega = \begin{bmatrix} 0.2 & 0.25 & 0.4 & 0.15 \\ 0.1 & 0.2 & 0.3 & 0.4 \\ 0.3 & 0.2 & 0.4 & 0.1 \\ 0.4 & 0.2 & 0.2 & 0.2 \end{bmatrix},$$

and $\mathcal{Q} = -0.36$, $\mathcal{S} = -2$, $\mathcal{R} = 2$.

By solving LMIs (5.60)–(5.62) via Matlab, we obtain that the controller gains are

$$\begin{aligned} K_{11} &= \begin{bmatrix} -27.5544 & -18.4128 \end{bmatrix}, & K_{21} &= \begin{bmatrix} -26.2919 & -17.100 \end{bmatrix}, \\ K_{31} &= \begin{bmatrix} -25.9322 & -16.9202 \end{bmatrix}, & K_{41} &= \begin{bmatrix} -26.5322 & -18.3330 \end{bmatrix}, \\ K_{12} &= \begin{bmatrix} -28.6199 & -18.3944 \end{bmatrix}, & K_{22} &= \begin{bmatrix} -25.8466 & -17.1084 \end{bmatrix}, \\ K_{32}^T &= \begin{bmatrix} -25.4786 & -16.8861 \end{bmatrix}, & K_{42} &= \begin{bmatrix} -29.4992 & -18.2711 \end{bmatrix}. \end{aligned}$$

In simulation, we assume that the initial condition is $x(0) = \begin{bmatrix} 0.3\pi & -0.1\pi \end{bmatrix}^T$, and the external disturbance is

$$w_k = \begin{cases} e^{-0.1(k-15)} \sin(0.3(k-15)), & 1 \le k \le 15, \\ 0, & 16 \le k \le 80. \end{cases}$$

Now, we adopt the state-feedback control to stabilize the open-loop fuzzy MJSs, shown in Fig. 5.2. From Fig. 5.3, it is easy to find out the state trajectories approximate to zero as time k goes by. By calculating, the optimal α^* for the strictly dissipative performance is 0.4696.

(1) Effects of Ω on the system performance: we investigate what impacts the different conditional probability matrices Ω have on the performance of systems for various

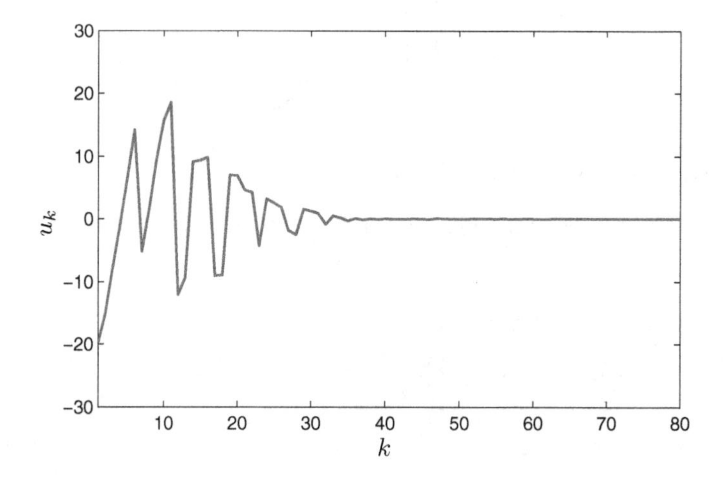

Fig. 5.2 Dissipative control

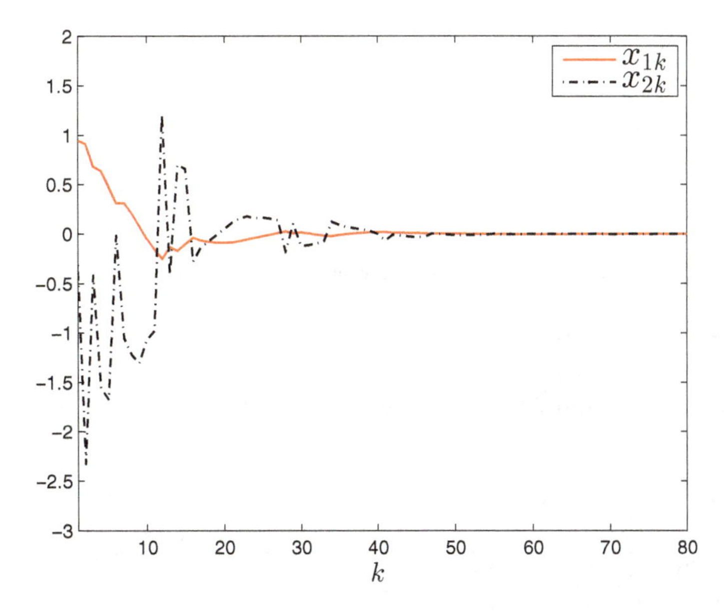

Fig. 5.3 State responses of the closed-loop system (5.37)

Table 5.5 Different values Ω

Case I (synchronous)	Case II (partially synchronous)
$\begin{bmatrix} 1 & 0 & 0 & 0 \\ 0 & 1 & 0 & 0 \\ 0 & 0 & 1 & 0 \\ 0 & 0 & 0 & 1 \end{bmatrix}$	$\begin{bmatrix} 1 & 0 & 0 & 0 \\ 0 & 1 & 0 & 0 \\ 0.3 & 0.2 & 0.4 & 0.1 \\ 0.4 & 0.2 & 0.2 & 0.2 \end{bmatrix}$
Case III (partially synchronous)	Case IV (asynchronous)
$\begin{bmatrix} 1 & 0 & 0 & 0 \\ 0.1 & 0.2 & 0.3 & 0.4 \\ 0.3 & 0.2 & 0.4 & 0.1 \\ 0.4 & 0.2 & 0.2 & 0.2 \end{bmatrix}$	$\begin{bmatrix} 0.2 & 0.25 & 0.4 & 0.15 \\ 0.1 & 0.2 & 0.3 & 0.4 \\ 0.3 & 0.2 & 0.4 & 0.1 \\ 0.4 & 0.2 & 0.2 & 0.2 \end{bmatrix}$

control issues, namely, dissipative, H_∞ and passive performance. The values Ω are shown in Table 5.5. Then, we acquire the optimal performance α^* by solving a set of LMIs in Theorem 5.7 together with adjusting parameter values $\mathcal{Q}$, $\mathcal{S}$ and $\mathcal{R}$, presented at Table 5.6. Based on Table 5.7, it is clear to find that the H_∞ performance and the passive performance have a similar trend: the higher the synchronization level is, the smaller the optimal performance α^* is. However, the tendency of the dissipative criterion is opposite.

Table 5.6 Values $\mathcal{Q}$, $\mathcal{S}$ and $\mathcal{R}$ for various control criteria

	$\mathcal{Q}$	$\mathcal{S}$	$\mathcal{R}$
Dissapativity	-0.36	-2	2
H_∞	-1	0	$\alpha^2+\alpha$
Passivity	0	1	2α

Table 5.7 Optimal performance α^* for various control criteria with different Ω

α^*	Case I	Case II	Case III	Case IV
Dissapativity	1.2625	0.9203	0.8939	0.4696
H_∞	0.1952	0.3024	0.3109	0.4536
Passivity	1.3797×10^{-14}	0.0089	0.0160	0.1350

Table 5.8 Optimal performance α^* for two different methods with different Ω

α^*	Case I	Case II	Case III	Case IV	Variables
Dissapativity	1.2625	0.9203	0.8939	0.4696	153
Alternative one	1.2574	0.8975	0.8690	0.3966	141

(2) Comparison between our approach and alternative one: the alternative approach is derived by choosing $\tilde{P}_{si}$ as $\tilde{P}_s$, that is, the Lyapunov function matrix becomes a fuzzy-basis-independent but mode-dependent one. From Table 5.8, it is evident that the number of variables of our technique needing solving is more than that of the alternative one. However, our method demonstrates a better performance than the alternation since α^* solved by our approach is bigger than that of alternative one. We employ the common Lyapunov function matrix $\tilde{P}_{si}$ as $\tilde{P}$ which has been used in [9]. Though this method has fewer unknown variables, there is no feasible result in Theorem 5.7. It is a trade off between computation complexity and conservatism. In this chapter, to obtain less conservatism for discrete-time fuzzy MJSs, we adopt the mode-dependent and fuzzy-basis-dependent Lyapunov function instead of only mode-dependent or common one.

5.7 Conclusion

In this chapter, the strictly $(\mathcal{Q}, \mathcal{S}, \mathcal{R})$-$\alpha$-dissipative asynchronous control problems have been investigated for both continuous-time and discrete-time T–S fuzzy systems with Markov jump. We have proposed sufficient conditions to guarantee that the closed-loop system is stochastically stable with $(\mathcal{Q}, \mathcal{S}, \mathcal{R})$-$\alpha$-dissipative performance via the Lyapunov function technique. The optimal dissipative performance and the corresponding asynchronous controller can be achieved via solving a convex optimization issue.

References

1. Wu, H.N.: Reliable robust H_∞ fuzzy control for uncertain nonlinear systems with Markovian jumping actuator faults. J. Dyn. Syst. Meas. Control **129**(3), 252–261 (2007)
2. Wu, H.-N., Cai, K.-Y.: Mode-independent robust stabilization for uncertain Markovian jump nonlinear systems via fuzzy control. IEEE Trans. Syst. Man Cybern. Part B (Cybern.) **36**(3), 509–519 (2005)
3. Tan, Z., Soh, Y., Xie, L.: Dissipative control for linear discrete-time systems. Automatica **35**(9), 1557–1564 (1999)
4. Willems, J.: Dissipative dynamical systems, part I: general theory. Arch. Ration. Mech. Anal. **45**(5), 321–393 (1972)
5. Willems, J.: Dissipative dynamical systems, part II: linear systems with quadratic supply rates. Arch. Ration. Mech. Anal. **45**(5), 321–393 (1972)
6. Wu, Z.-G., Shi, P., Shu, Z., Su, H., Lu, R.: Passivity-based asynchronous control for Markov jump systems. IEEE Trans. Autom. Control **62**(4), 2020–2025 (2017)
7. Tian, E., Yue, D., Yang, T.C., Gu, Z., Lu, G.: T-S fuzzy model-based robust stabilization for networked control systems with probabilistic sensor and actuator failure. IEEE Trans. Fuzzy Syst. **19**(19), 553–561 (2011)
8. Tanaka, K., Ikeda, T., Wang, H.O.: Robust stability of a class of uncertain nonlinear systems via fuzzy control: quadratic stability, H_∞ control theory, and linear matrix inequalities. IEEE Trans. Fuzzy Syst. **40**(1), 1–13 (1996)
9. Wu, Z.-G., Shi, P., Su, H., Lu, R.: Dissipativity-based sampled-data fuzzy control design and its application to truck-trailer system. IEEE Trans. Fuzzy Syst. **23**(5), 1669–1679 (2015)

Chapter 6
Extended Dissipativity-Based Control for Fuzzy Switched Systems with Intermittent Measurements

6.1 Introduction

Because of sensor constraints, we may be unable to directly measure the state information. Thus, the state feedback control may lose its function [1]. Instead, the work in [2] has designed the output feedback controller without using mode information to address the infinite-horizon optimal control problem for MJSs. In this chapter, we investigate the asynchronous output feedback control problem for networked fuzzy switched systems with a desired extended dissipative performance by adopting the HMM principle. In practice, the communication channels in networked control systems are often imperfect, where data packets drop intermittently and inevitably [3]. We apply a stochastic Bernoulli process to model this random phenomenon [4]. The Lyapunov function is adopted to obtain sufficient conditions such that the closed-loop system is stochastically stable with a satisfactory extended dissipativity. The desired asynchronous output feedback controller gains can be achieved by solving a set of matrix inequalities.

6.2 Preliminary Analysis

Consider the following discrete-time T–S fuzzy MJSs:
Plant rule i^{δ_k}: IF θ_{1k} is $\zeta_{i1}^{\delta_k}$, θ_{2k} is $\zeta_{i2}^{\delta_k}$, ..., and θ_{lk} is $\zeta_{il}^{\delta_k}$, THEN

$$\begin{cases} x_{k+1} = A_{i\delta_k} x_k + B_{i\delta_k} u_k + C_{i\delta_k} w_k, \\ y_k = D_{i\delta_k} x_k, \\ z_k = E_{i\delta_k} x_k + F_{i\delta_k} w_k, \end{cases} \tag{6.1}$$

where θ_{jk} $(j \in \{1, 2, \ldots, l\})$ is the premise variable and $\zeta_{ij}^{\delta_k}$ is the fuzzy set $(i \in \mathcal{R} = \{1, 2, \ldots, r\})$. $x_k \in R^{n_x}$, $u_k \in R^{n_u}$, $y_k \in R^{n_y}$ and $z_k \in R^{n_z}$ are the state,

S. Dong et al., *Control and Filtering of Fuzzy Systems with Switched Parameters*, Studies in Systems, Decision and Control 268,
https://doi.org/10.1007/978-3-030-35566-1_6

the control input, the measurable output and the output performance, respectively. $w_k \in R^{n_w}$ that belongs to $l_2[0, +\infty)$ is the external noise. System matrices $A_{i\delta_k}$, $B_{i\delta_k}$, $C_{i\delta_k}$, $D_{i\delta_k}$, $E_{i\delta_k}$ and $F_{i\delta_k}$ are known with appropriate dimensions, where the random variable δ_k ($\delta_k \in \mathcal{N} = \{1, 2, \ldots, N\}$) is introduced to describe the Markov jump phenomenon with the transition probability matrix $\Pi = [\phi_{mn}]$ and

$$\Pr\{\delta_{k+1} = n|\delta_k = m\} = \phi_{mn}, \ m, \ n \in \mathcal{N}, \tag{6.2}$$

where $\phi_{mn} \geq 0$ and $\sum_{n=1}^{N} \phi_{mn} = 1$.

According to (6.1), by assuming $\delta_k = m$, we have the following systems in a compact form:

$$\begin{cases} x_{k+1} = A_{hm} x_k + B_{hm} u_k + C_{hm} w_k, \\ y_k = D_{hm} x_k, \\ z_k = E_{hm} x_k + F_{hm} w_k, \end{cases} \tag{6.3}$$

where

$$A_{hm} = \sum_{i=1}^{r} h_{im} A_{im}, \ B_{hm} = \sum_{i=1}^{r} h_{im} B_{im},$$
$$C_{hm} = \sum_{i=1}^{r} h_{im} C_{im}, \ D_{hm} = \sum_{i=1}^{r} h_{im} D_{im},$$
$$E_{hm} = \sum_{i=1}^{r} h_{im} E_{im}, \ F_{hm} = \sum_{i=1}^{r} h_{im} F_{im}.$$

h_{im} is the abbreviation of $h_{im}(\theta_k)$ ($\theta_k = \left[\theta_{1k}, \theta_{2k}, \ldots, \theta_{lk}\right]$), which is the fuzzy basis function with $h_{im} = h_{im}(\theta_k) = \frac{\prod_{j=1}^{l} \zeta_{ij}^{m}(\theta_{jk})}{\sum_{i=1}^{r} \prod_{j=1}^{l} \zeta_{ij}^{m}(\theta_{jk})}$. $\zeta_{ij}^{m}(\theta_{jk})$ is the grade of membership of θ_{jk} in ζ_{ij}^{m}. It is observed that $\sum_{i=1}^{r} h_{im} = 1$ and $h_{im} \geq 0$.

In practice, due to network-induced limitations, data dropouts happen unavoidably, which means that the output signal y_k is measured intermittently. In this situation, a Bernoulli process is employed to model this random phenomenon as follows:

$$\hat{y}_k = \alpha_k y_k, \tag{6.4}$$

where the variable α_k satisfies the Bernoulli distribution with $\Pr\{\alpha_k = 1\} = \alpha$ ($a \in (0, 1]$) and $\Pr\{\alpha_k = 0\} = 1 - \alpha$.

Considering the unavailability of state signals, we apply output measurements to design the following asynchronous output feedback controller by combining the HMM property:

$$u_k = K_{\eta_k} \hat{y}_k, \tag{6.5}$$

where K_{η_k} is the controller gain to be solved. The variable η_k ($\eta_k \in \mathcal{G} = \{1, 2, \ldots, G\}$) is adopted to observe the switched mode δ_k. Then the original systems (6.3) and the designed controller (6.5) may run asynchronously. η_k is subject to the conditional transition probability matrix $\Psi = [\rho_{mg}]$ with

$$\Pr\{\eta_k = g|\delta_k = m\} = \rho_{mg}, \quad m \in \mathcal{N}, \tag{6.6}$$

where $\rho_{mg} \in [0, 1]$, and $\sum_{g=1}^{G} \rho_{mg} = 1$.

Remark 6.1 For the fuzzy control problem, many works have been published, such as the fuzzy logic control for multi input-multi output nonlinear processes [5], the multi-level adaptive control for actuated microplates [6] and the fuzzy state feedback control by using the PDC approach [7]. It is worth pointing out that the PDC method is classical and used widely including the fuzzy state feedback control for the robotic arm [8] and the optimal state feedback fuzzy control via swarm intelligence optimization algorithms [9]. In this chapter, owing to unavailability of states, output information is used to investigate the control problem for T–S fuzzy switched systems. On the other hand, since premise variables and the fuzzy set could depend on states, the PDC method of designing fuzzy controllers is not applicable any more [10–12]. Further, considering that the switched mode δ_k cannot be directly used, the HMM is exploited to detect the mode information by a stochastic variable η_k. Hence, the static asynchronous output feedback controller without using the PDC approach is designed, shown in (6.5).

Recalling (6.3), (6.4) and (6.5) with $\eta_k = g$, we have the following closed-loop systems:

$$\begin{cases} x_{k+1} = \bar{A}_{hmg} x_k + C_{hm} w_k, \\ z_k = E_{hm} x_k + F_{hm} w_k, \end{cases} \tag{6.7}$$

where

$$\bar{A}_{hmg} = A_{hm} + \alpha_k B_{hm} K_g D_{hm} = \sum_{i=1}^{r} \sum_{j=1}^{r} h_{im} h_{jm} \bar{A}_{ijmg},$$

$$\bar{A}_{ijmg} = A_{im} + \alpha_k B_{im} K_g D_{jm}.$$

Remark 6.2 In dealing with the output feedback control problem, the closed-loop system is often reformulated as a descriptor system by using the system-augmentation method, which increases the dimension of studied systems [11]. On the other hand, some published works put some constraints on system matrices [13], which limit the application of the designed output feedback controller to some extent. In this chapter, neither is adopted.

In the following, a lemma, an assumption and a definition are introduced, which are helpful for later investigation.

Lemma 6.3 ([14]) *If there exist a scalar ε, matrices $\mathcal{T}$, $\mathcal{X}$, $\mathcal{A}$ and $\mathcal{B}$ satisfying*

$$\begin{bmatrix} \mathcal{T} & * \\ \mathcal{A}^T + \varepsilon\mathcal{B} & -\varepsilon\mathcal{X} - \varepsilon\mathcal{X}^T \end{bmatrix} < 0, \tag{6.8}$$

we have

$$\mathcal{T} + \mathcal{A}\mathcal{X}^{-1}\mathcal{B} + \mathcal{B}^T(\mathcal{X}^{-1})^T\mathcal{A}^T < 0. \tag{6.9}$$

Assumption 6.1 ([15]) For known real matrices $U_1 \leq 0, U_2, U_3 = U_3^T$ and $U_4 \geq 0$, the following conditions are satisfied:
(1) $||F_{hm}||||U_4|| = 0$;
(2) $(||U_1|| + ||U_2||)||U_4|| = 0$;
(3) $F_{hm}^T U_1 F_{hm} + F_{hm}^T U_2 + U_2^T F_{hm} + U_3 > 0$.

Definition 6.1 ([15]) For known real matrices $U_1 = -(U_1^+)^T U_1^+ \leq 0,\ U_2,\ U_3 = U_3^T$ and $U_4 = (U_4^+)^T U_4^+ \geq 0$ with Assumption 6.1, system (6.7) is said to be extended dissipative if the following inequality holds for any positive integer k_f and $a \in [0, 1, 2, \ldots, k_f]$ under the zero initial condition:

$$E\left\{\sum_{a=0}^{k_f} J(a)\right\} > \sup_{0 \leq a \leq k_f} E\{z_a^T U_4 Z_a\}, \tag{6.10}$$

where

$$J(a) = z_a^T U_1 z_a + 2z_a^T U_2 w_a + w_a^T U_3 w_a.$$

Remark 6.4 The extended dissipative performance was firstly introduced in [15] to investigate the synchronous and mode-independent filter problems for MJSs with time-varying delays. The new performance includes some widely used performance indexes. For instance, by letting $U_1 = -I,\ U_2 = 0,\ U_3 = \gamma^2 I$ and $U_4 = 0$, (6.10) is changed into H_∞ control. The strict dissipative performance index is achieved if $U_3 = R - \gamma I\ (\gamma > 0)$ and $U_4 = 0$. We have the $L_2 - L_\infty$ (energy-to-peak) performance by assuming $U_1 = 0,\ U_2 = 0,\ U_3 = \gamma^2 I$ and $U_4 = I$. Further, if z_k and w_k have the same dimension, that is, $n_z = n_w$, the passivity will be obtained with $U_1 = 0,\ U_2 = I,\ U_3 = \gamma I$ and $U_4 = 0$. The chapter adopts the extended dissipative performance to study the asynchronous output feedback control problem for fuzzy switched systems.

The problem of this chapter is reformulated as follows: devise a suitable asynchronous output feedback controller in the form of (6.5) such that
(1) When $w_k \equiv 0$, the closed-loop system (6.7) is stochastically stable, namely, $E\left\{\sum_{k=0}^{\infty} |x_k|^2 | x_0, \delta_0\right\} < \infty$;
(2) Based on Definition 6.1, system (6.7) achieves a desired extended dissipative performance.

6.3 Main Results

In this section, we first present a sufficient condition to guarantee the stochastic stability of the closed-loop system (6.7) with a satisfactory extended dissipative performance. Then, we focus on developing a solution to the asynchronous output feedback controller (6.5).

Theorem 6.5 *For given matrices U_1, U_2, U_3 and U_4 satisfying Definition 6.1, system (6.7) is stochastically stable with a desired extended dissipativity if there exist matrices $P_m > 0$ and $Q_{mg} > 0$ for any $m \in \mathcal{N}$ and $g \in \mathcal{G}$ satisfying*

$$\sum_{g=1}^{G} \rho_{mg} Q_{mg} < P_m, \tag{6.11}$$

$$\Gamma_{hmg} < 0, \tag{6.12}$$

$$\begin{bmatrix} -P_m & * \\ U_4^+ E_{hm} & -I \end{bmatrix} < 0, \tag{6.13}$$

where

$$\Gamma_{hmg} = \begin{bmatrix} -Q_{mg} & * & * & * & * \\ -U_2^T E_{hm} & \Gamma_{hm}^{22} & * & * & * \\ U_1^+ E_{hm} & U_1^+ F_{hm} & -I & * & * \\ \Gamma_{hmg}^{41} & \Gamma_{hmg}^{42} & 0 & -\Gamma^{44} & * \\ \Gamma_{hmg}^{51} & 0 & 0 & 0 & -\Gamma^{55} \end{bmatrix},$$

$$\Gamma_{hm}^{22} = -F_{hm}^T U_2 - U_2^T F_{hm} - U_3,$$
$$\Gamma_{hmg}^{41} = [\sqrt{\phi_{m1}}(P_1 \tilde{A}_{hmg})^T \cdots \sqrt{\phi_{mN}}(P_N \tilde{A}_{hmg})^T]^T,$$
$$\Gamma_{hmg}^{42} = [\sqrt{\phi_{m1}}(P_1 C_{hm})^T \cdots \sqrt{\phi_{mN}}(P_N C_{hm})^T]^T,$$
$$\Gamma_{hmg}^{51} = [\sqrt{\phi_{m1}}(P_1 \tilde{B}_{hmg})^T \cdots \sqrt{\phi_{mN}}(P_N \tilde{B}_{hmg})^T]^T,$$
$$\Gamma^{55} = \Gamma^{44} = diag\{P_1, \ldots, P_N\},$$
$$\tilde{A}_{hmg} = A_{hm} + \alpha B_{hm} K_g D_{hm},$$
$$\tilde{B}_{hmg} = f B_{hm} K_g D_{hm}, \; f = \sqrt{\alpha - \alpha^2}.$$

Proof Applying Schur Complement to (6.12) and (6.13), respectively, we have

$$\begin{aligned} &\begin{bmatrix} -Q_{mg} & * \\ -U_2^T E_{hm} & \Gamma_{hm}^{22} \end{bmatrix} - \begin{bmatrix} E_{hm}^T \\ F_{hm}^T \end{bmatrix} U_1 \begin{bmatrix} E_{hm}^T \\ F_{hm}^T \end{bmatrix}^T \\ &+ \sum_{n=1}^{N} \phi_{mn} \left(\begin{bmatrix} \tilde{A}_{hmg}^1 & * \\ C_{hm}^T P_n \tilde{A}_{hmg} & C_{hm}^T P_n C_{hm} \end{bmatrix} \right) < 0 \end{aligned} \tag{6.14}$$

and

$$E_{hm}^T U_4 E_{hm} < P_m, \tag{6.15}$$

where

$$\tilde{A}_{hmg}^1 = \tilde{A}_{hmg}^T P_n \tilde{A}_{hmg} + \tilde{B}_{hmg}^T P_n \tilde{B}_{hmg}.$$

Further, it easily follows from (6.11) that

$$\begin{aligned} \Lambda_{hm} =& \sum_{g=1}^{G} \rho_{mg} \sum_{n=1}^{N} \phi_{mn} \left(\begin{bmatrix} \tilde{A}_{hmg}^1 & * \\ C_{hm}^T P_n \tilde{A}_{hmg} & C_{hm}^T P_n C_{hm} \end{bmatrix} \right) \\ &+ \begin{bmatrix} -P_m & * \\ -U_2^T E_{hm} & \Gamma_{hm}^{22} \end{bmatrix} - \begin{bmatrix} E_{hm}^T \\ F_{hm}^T \end{bmatrix} U_1 \begin{bmatrix} E_{hm}^T \\ F_{hm}^T \end{bmatrix}^T < 0 \end{aligned} \tag{6.16}$$

and

$$\sum_{g=1}^{G} \rho_{mg} \sum_{n=1}^{N} \phi_{mn} \tilde{A}_{hmg}^1 - P_m < 0. \tag{6.17}$$

We design the following Lyapunov function:

$$V_k = x_k^T P_{\delta_k} x_k, \tag{6.18}$$

where $P_{\delta_k} > 0$. It is assumed that $\delta_k = m$ and $\delta_{k+1} = n$. Then computing the difference of V_k and taking the expectation with $w_k \equiv 0$, we have

$$\begin{aligned} E\{\Delta V_k\} &= E\{x_{k+1}^T P_n x_{k+1}\} - E\{x_k^T P_m x_k\} \\ &= x_k^T \left(\sum_{g=1}^{G} \rho_{mg} \sum_{n=1}^{N} \phi_{mn} \tilde{A}_{hmg}^1 - P_m \right) x_k < 0, \end{aligned} \tag{6.19}$$

which means that system (6.7) is stochastically stable.

Considering the extended dissipative performance in (6.10), we obtain

$$\begin{aligned} &E\{\Delta V_k - z_k^T U_1 z_k - 2 z_k^T U_2 w_k - w_k^T U_3 w_k\} \\ &= \begin{bmatrix} x_k^T & w_k^T \end{bmatrix} \Lambda_{hm} \begin{bmatrix} x_k^T & w_k^T \end{bmatrix}^T < 0. \end{aligned} \tag{6.20}$$

In the following, the main work is to validate that (6.10) holds under Assumption 6.1.

(1) When $U_4 = 0$, it readily follows from (6.20) that

$$\sum_{a=0}^{k_f} E\{\Delta V_a - z_a^T U_1 z_a - 2 z_a^T U_2 w_a - w_a^T U_3 w_a\} < 0. \tag{6.21}$$

Subsequently, we clearly find that

$$E\left\{\sum_{a=0}^{k_f} J(a)\right\} > E\{V_{k_f+1}\} - E\{V_0\} = E\{V_{k_f+1}\} > 0. \tag{6.22}$$

Based on Assumption 6.1 with $U_4 = 0$, it derives that (6.10) holds.

(2) When $U_4 \neq 0$, it follows from Assumption 6.1 that $F_{hm} = 0$, $U_1 = 0$, $U_2 = 0$ and $U_3 > 0$. The following inequality is obtained:

$$E\left\{\sum_{a=0}^{k_f-1} J(a)\right\} > E\{V_{k_f}\} = E\{x_{k_f}^T P_{m^{k_f}} x_{k_f}\} > 0, \tag{6.23}$$

where m^{k_f} is the jump mode at time k_f. Then from (6.15), we have

$$E_{hm}^T U_4 E_{hm} < P_m. \tag{6.24}$$

In the case of $k = k_f$ and $m = m^{k_f}$, it further follows that

$$E\left\{\sum_{a=0}^{k_f} J(a)\right\} > E\{x_{k_f}^T P_{m^{k_f}} x_{k_f}\} > E\{z_{k_f}^T U_4 z_{k_f}\}. \tag{6.25}$$

Due to $U_1 = 0$, $U_2 = 0$ and $U_3 > 0$, we have

$$E\left\{\sum_{a=0}^{k_f} J(a)\right\} > \sup_{0\le a\le k_f} E\{z^T(a) U_4 Z(a)\}.$$

Hence, we can clearly conclude that system (6.7) is stochastically stable with a desired extended dissipative performance. The proof is completed.

Based on Theorem 6.5, an algorithm to solve the asynchronous output feedback controller gain in (6.5) is developed, as follows.

Theorem 6.6 *For given matrices U_1, U_2, U_3 and U_4 satisfying Definition 6.1, system (6.7) is stochastically stable with a desired extended dissipativity if there exist matrices $P_m > 0$, $Q_{mg} > 0$, Y_g, V_g and a scalar ε for any $m \in \mathcal{N}$, $g \in \mathcal{G}$ and any $i, j \in \mathcal{R}$ subject to*

$$\sum_{g=1}^{G} \rho_{mg} Q_{mg} < P_m, \tag{6.26}$$

$$\Upsilon_{iimg} < 0, \tag{6.27}$$

$$\Upsilon_{ijmg} + \Upsilon_{jimg} < 0,\ j < i, \tag{6.28}$$

$$\begin{bmatrix} -P_m & * \\ U_4^+ E_{im} & -I \end{bmatrix} < 0, \tag{6.29}$$

where

$$\Upsilon_{ijmg} = \begin{bmatrix} -Q_{mg} & * & * & * & * & * \\ -U_2^T E_{im} & \Upsilon_{im}^{22} & * & * & * & * \\ U_1^+ E_{im} & U_1^+ F_{im} & -I & * & * & * \\ \Upsilon_{ijmg}^{41} & \Upsilon_{im}^{42} & 0 & -\Upsilon^{44} & * & * \\ \Upsilon_{ijmg}^{51} & 0 & 0 & 0 & -\Upsilon^{55} & * \\ \varepsilon Y_g D_{jm} & 0 & 0 & \Upsilon_{img}^{64} & \Upsilon_{img}^{65} & \Upsilon_g^{66} \end{bmatrix},$$

$$\Upsilon_{im}^{22} = -F_{im}^T U_2 - U_2^T F_{im} - U_3,$$

$$\Upsilon_{ijmg}^{41} = [\sqrt{\phi_{m1}}(P_1 A_{im} + \alpha B_{im} Y_g D_{jm})^T \\ \cdots \sqrt{\phi_{mN}}(P_N A_{im} + \alpha B_{im} Y_g D_{jm})^T]^T,$$

$$\Upsilon_{im}^{42} = \left[\sqrt{\phi_{m1}}(P_1 C_{im})^T \ \cdots \ \sqrt{\phi_{mN}}(P_N C_{im})^T\right]^T,$$

$$\Upsilon_{ijmg}^{51} = [\sqrt{\phi_{m1}} f(B_{im} Y_g D_{jm})^T \cdots \sqrt{\phi_{mN}} f(B_{im} Y_g D_{jm})^T]^T,$$

$$\Upsilon^{55} = \Upsilon^{44} = diag\{P_1, \ldots, P_N\},$$

$$\Upsilon_{img}^{64} = [\sqrt{\phi_{m1}}\alpha(P_1 B_{im} - B_{im} V_g)^T \ \cdots \sqrt{\phi_{mN}}\alpha(P_N B_{im} - B_{im} V_g)^T],$$

$$\Upsilon_{img}^{65} = [\sqrt{\phi_{m1}} f(P_1 B_{im} - B_{im} V_g)^T \ \cdots \sqrt{\phi_{mN}} f(P_N B_{im} - B_{im} V_g)^T],$$

$$\Upsilon_g^{66} = -\varepsilon V_g - \varepsilon V_g^T.$$

Moreover, the asynchronous output feedback controller gains can be solved from

$$K_g = V_g^{-1} Y_g. \tag{6.30}$$

Proof From (6.27), (6.28), we have

$$\begin{aligned} \Upsilon_{hmg} &= \sum_{i=1}^{r}\sum_{j=1}^{r} h_{im} h_{jm} \Upsilon_{ijmg} \\ &= \sum_{i=1}^{r} h_{im}^2 \Upsilon_{iimg} + \sum_{i=1}^{r}\sum_{j<i}^{r} h_{im} h_{jm} (\Upsilon_{ijmg} + \Upsilon_{jimg}) < 0 \end{aligned} \tag{6.31}$$

where

$$\Upsilon_{hmg} = \begin{bmatrix} -Q_{hmg} & * & * & * & * & * \\ -U_2^T E_{hm} & \Gamma_{hm}^{22} & * & * & * & * \\ U_1^+ E_{hm} & U_1^+ F_{hm} & -I & * & * & * \\ \Upsilon_{hmg}^{41} & \Upsilon_{hm}^{42} & 0 & -\Upsilon^{44} & * & * \\ \Upsilon_{hmg}^{51} & 0 & 0 & 0 & -\Upsilon^{55} & * \\ \varepsilon Y_g D_{hm} & 0 & 0 & \Upsilon_{hmg}^{64} & \Upsilon_{hmg}^{65} & \Upsilon_g^{66} \end{bmatrix},$$

$$\Upsilon_{hmg}^{41} = [\sqrt{\phi_{m1}}(P_1 A_{hm} + \alpha B_{hm} Y_g D_{hm})^T \\ \cdots \sqrt{\phi_{mN}}(P_N A_{hm} + \alpha B_{hm} Y_g D_{hm})^T]^T,$$

$$\Upsilon_{hm}^{42} = \left[\sqrt{\phi_{m1}}(P_1 C_{hm})^T \ \cdots \ \sqrt{\phi_{mN}}(P_N C_{hm})^T\right]^T,$$

$$\Upsilon_{hmg}^{51} = [\sqrt{\phi_{m1}} f(B_{hm} Y_g D_{hm})^T \ \cdots \sqrt{\phi_{mN}} f(B_{hm} Y_g D_{hm})^T]^T,$$

$$\Upsilon^{55} = \Upsilon^{44} = \text{diag}\{P_1, \ldots, P_N\},$$

$$\Upsilon_{hmg}^{64} = [\sqrt{\phi_{m1}}\alpha(P_1 B_{hm} - B_{hm} V_g)^T \ \cdots \sqrt{\phi_{mN}}\alpha(P_N B_{hm} - B_{hm} V_g)^T],$$

$$\Upsilon_{hmg}^{65} = [\sqrt{\phi_{m1}} f(P_1 B_{hm} - B_{hm} V_g)^T \ \cdots \sqrt{\phi_{mN}} f(P_N B_{hm} - B_{hm} V_g)^T],$$

$$\Upsilon_g^{66} = -\varepsilon V_g - \varepsilon V_g^T.$$

Based on Lemma 6.3, we clearly find that (6.31) can guarantee the correctness of (6.12) if $Y_g = V_g K_g$. Moreover, it is clearly seen that (6.26) and (6.29) can ensure that (6.11) and (6.13) hold, respectively. The proof is completed.

6.4 Extension to Fuzzy Switched Systems with Sojourn Probabilities

Inspired by above analysis, we now extend our study to fuzzy switched systems with sojourn probabilities. That is, the probability that the systems stay in each subsystem is supposed to be given with $\Pr\{\delta_k = m\} = \Pr\{\delta_k = m|\delta_{k-1} = b\} = \phi_{bm} = \bar{\beta}_m$ $(m, \ b \in \mathcal{N})$ and $\sum_{m=1}^{N} \bar{\beta}_m = 1$. Further, a random variable $\beta_m(k)$ is defined as

$$\beta_m(k) = \begin{cases} 1, & \delta_k = m, \\ 0, & \delta_k \neq m, \end{cases} \tag{6.32}$$

and $\sum_{m=1}^{N} \beta_m(k) = 1$, which implies that every time, only one switched mode works. Then the defuzzied systems from (6.1) are inferred as

$$\begin{cases} x_{k+1} = \sum\limits_{m=1}^{N} \beta_m(k)[A_{hm} x_k + B_{hm} u_k + C_{hm} w_k], \\ y_k = \sum\limits_{m=1}^{N} \beta_m(k) D_{hm} x_k, \\ z_k = \sum\limits_{m=1}^{N} \beta_m(k)[E_{hm} x_k + F_{hm} w_k]. \end{cases} \tag{6.33}$$

Next, by applying the asynchronous output feedback control design in (6.5), we obtain the following dynamics of closed-loop systems:

$$\begin{cases} x_{k+1} = \sum_{m=1}^{N} \beta_m(k)[\bar{A}_{hmg}x_k + C_{hm}w_k], \\ z_k = \sum_{m=1}^{N} \beta_m(k)[E_{hm}x_k + F_{hm}w_k], \end{cases} \tag{6.34}$$

where

$$\bar{A}_{hmg} = A_{hm} + \alpha_k B_{hm} K_g D_{hm} = \sum_{i=1}^{r}\sum_{j=1}^{r} h_{im}h_{jm}\bar{A}_{ijmg},$$
$$\bar{A}_{ijmg} = A_{im} + \alpha_k B_{im} K_g D_{jm}.$$

Remark 6.7 In contrast with N^2 transition probabilities in fuzzy MJSs, there are only N sojourn probabilities to be known before stochastic systems work, which is easier to measure with less computation. This directly motivates us to investigate the asynchronous output feedback control problem for fuzzy switched systems with sojourn probabilities. It is worth pointing out that there is only one mode operating every time, which implies that if $m \neq n$, $E\{\beta_m(k)\beta_n(k)\} = 0$; otherwise, $E\{\beta_m(k)\beta_m(k)\} = \bar{\beta}_m$.

In the following, a sufficient condition is firstly given for stochastic stability with a desired extended dissipative performance for fuzzy switched systems (6.34). Then, the corresponding solution to asynchronous output feedback controller gains is developed.

Theorem 6.8 *For given matrices U_1, U_2, U_3 and U_4 satisfying Definition 6.1, system (6.34) is stochastically stable with a desired extended dissipative performance if there exist matrices $P > 0$ and $Q_{mg} > 0$ for any $m \in \mathcal{N}$ and $g \in \mathcal{G}$ satisfying*

$$\sum_{m=1}^{N}\sum_{g=1}^{G} \bar{\beta}_m \rho_{mg} Q_{mg} < P, \tag{6.35}$$

$$\Theta_{hmg} < 0, \tag{6.36}$$

$$\begin{bmatrix} -P & * \\ \Omega_h^{21} & \Omega^{22} \end{bmatrix} < 0, \tag{6.37}$$

where

$$\Theta_{hmg} = \begin{bmatrix} -Q_{mg} & * & * & * & * \\ -U_2^T E_{hm} & \Theta_{hm}^{22} & * & * & * \\ U_1^+ E_{hm} & U_1^+ F_{hm} & -I & * & * \\ \Theta_{hmg}^{41} & \Theta_{hm}^{42} & 0 & -P & * \\ \Theta_{hmg}^{51} & 0 & 0 & 0 & -P \end{bmatrix},$$

$$\Theta_{hm}^{22} = -F_{hm}^T U_2 - U_2^T F_{hm} - U_3, \ \Theta_{hm}^{42} = PC_{hm},$$

$$\Theta_{hmg}^{41} = P\tilde{A}_{hmg}, \ \Theta_{hmg}^{51} = P\tilde{B}_{hmg},$$

$$\Omega_h^{21} = [\sqrt{\bar{\beta}_1} E_{h1}^T, \ldots, \sqrt{\bar{\beta}_N} E_{hN}^T]^T,$$

$$\Omega^{22} = -diag\{U_4^{-1}, \ldots, U_4^{-1}\},$$

$$\tilde{A}_{hmg} = A_{hm} + \alpha B_{hm} K_g D_{hm},$$

$$\tilde{B}_{hmg} = f B_{hm} K_g D_{hm}, \ f = \sqrt{\alpha - \alpha^2}.$$

Proof Applying Schur Complement to (6.36) and (6.37), respectively, we have

$$\begin{bmatrix} -Q_{mg} & * \\ -U_2^T E_{hm} & \Theta_{hm}^{22} \end{bmatrix} - \begin{bmatrix} E_{hm}^T \\ F_{hm}^T \end{bmatrix} U_1 \begin{bmatrix} E_{hm}^T \\ F_{hm}^T \end{bmatrix}^T + \begin{bmatrix} \tilde{\mathcal{A}}_{hmg}^1 & * \\ C_{hm}^T P \tilde{A}_{hmg} & C_{hm}^T P C_{hm} \end{bmatrix} < 0, \tag{6.38}$$

and

$$\sum_{m=1}^{N} \bar{\beta}_m E_{hm}^T U_4 E_{hm} < P, \tag{6.39}$$

where

$$\tilde{\mathcal{A}}_{hmg}^1 = \tilde{A}_{hmg}^T P \tilde{A}_{hmg} + \tilde{B}_{hmg}^T P \tilde{B}_{hmg}.$$

From (6.35), it follows that

$$\begin{aligned} \Lambda_{hm}^1 = & \sum_{m=1}^{N} \sum_{g=1}^{G} \bar{\beta}_m \rho_{mg} \left(\begin{bmatrix} \tilde{\mathcal{A}}_{hmg}^1 & * \\ C_{hm}^T P \tilde{A}_{hmg} & C_{hm}^T P C_{hm} \end{bmatrix} \right. \\ & \left. + \begin{bmatrix} -P & * \\ -U_2^T E_{hm} & \Theta_{hm}^{22} \end{bmatrix} - \begin{bmatrix} E_{hm}^T \\ F_{hm}^T \end{bmatrix} U_1 \begin{bmatrix} E_{hm}^T \\ F_{hm}^T \end{bmatrix}^T \right) < 0 \end{aligned} \tag{6.40}$$

and

$$\sum_{m=1}^{N} \sum_{g=1}^{G} \bar{\beta}_m \rho_{mg} \tilde{\mathcal{A}}_{hmg}^1 - P < 0. \tag{6.41}$$

The candidate Lyapunov function is chosen as

$$V_k = x_k^T P x_k. \tag{6.42}$$

On the other hand, $E\{\beta_m(k)\beta_n(k)\} = 0$ holds if $m \neq n$. Otherwise, $E\{\beta_m(k)\beta_m(k)\} = \bar{\beta}_m$. In the case of $w_k \equiv 0$, we have

$$\begin{aligned} E\{\Delta V_k\} &= E\{x_{k+1}^T P x_{k+1}\} - E\{x_k^T P x_k\} \\ &= x_k^T \left(\sum_{m=1}^{N} \sum_{g=1}^{G} \bar{\beta}_m \rho_{mg} \tilde{A}_{hmg}^1 - P \right) x_k < 0. \end{aligned} \tag{6.43}$$

Hence, we can readily observe that system (6.34) is stochastically stable. Since the remaining proof of the extended dissipative performance is similar to that in Theorem 6.5, we omit it. The proof is completed.

Theorem 6.9 *For given matrices U_1, U_2, U_3 and U_4 satisfying Definition 6.1, system (6.34) is stochastically stable with a desired extended dissipativity performance if there exist matrices $P > 0$, $Q_{mg} > 0$, Y_g, V_g and a scalar ε for any $m \in \mathcal{N}$, $g \in \mathcal{G}$ and any $i, j \in \mathcal{R}$ subject to*

$$\sum_{m=1}^{N} \sum_{g=1}^{G} \bar{\beta}_m \rho_{mg} Q_{mg} < P, \tag{6.44}$$

$$\Xi_{iimg} < 0, \tag{6.45}$$

$$\Xi_{ijmg} + \Xi_{jimg} < 0, \ j < i, \tag{6.46}$$

$$\begin{bmatrix} -P & * \\ \Sigma_i^{21} & \Sigma^{22} \end{bmatrix} < 0, \tag{6.47}$$

where

$$\Xi_{ijmg} = \begin{bmatrix} -Q_{mg} & * & * & * & * & * \\ -U_2^T E_{im} & \Xi_{im}^{22} & * & * & * & * \\ U_1^+ E_{im} & U_1^+ F_{im} & -I & * & * & * \\ \Xi_{ijmg}^{41} & \Xi_{im}^{42} & 0 & -P & * & * \\ \Xi_{ijmg}^{51} & 0 & 0 & 0 & -P & * \\ \varepsilon Y_g D_{jm} & 0 & 0 & \Xi_{img}^{64} & \Xi_{img}^{65} & \Xi_g^{66} \end{bmatrix},$$

$$\Xi_{im}^{22} = -F_{im}^T U_2 - U_2^T F_{im} - U_3,$$

$$\Xi_{ijmg}^{41} = P A_{im} + \alpha B_{im} Y_g D_{jm}, \ \Xi_{im}^{42} = P C_{im},$$

$$\Xi_{ijmg}^{51} = f B_{im} Y_g D_{jm}, \ \Xi_{img}^{64} = \alpha (P B_{im} - B_{im} V_g)^T,$$

$$\Xi_{img}^{65} = f (P B_{im} - B_{im} V_g)^T, \ \Xi_g^{66} = -\varepsilon V_g - \varepsilon V_g^T,$$

$$\Sigma_i^{21} = [\sqrt{\bar{\beta}_1} E_{i1}^T, \ldots, \sqrt{\bar{\beta}_N} E_{iN}^T]^T, \ \Sigma^{22} = -diag\{U_4^{-1}, \ldots, U_4^{-1}\}.$$

Moreover, the asynchronous output feedback controller gains can be solved from

$$K_g = V_g^{-1} Y_g. \tag{6.48}$$

Proof Adopting the similar proof process in Theorem 6.6, we can derive Theorem 6.9. The proof is completed.

Remark 6.10 According to Theorem 6.5, it is clearly observed that the nonlinear coupling exists among matrices P_m, K_g and D_{hm}. To decrease the computational difficulty, we transform the nonlinear matrix coupling to the nonlinear coupling between a scalar ε and matrix V_g by introducing a slack matrix Q_{mg} and adopting Lemma 6.3, shown in Theorem 6.6. Then we can adopt the following two steps to solve the controller gains:

Step 1: Choose an appropriate scalar ε to ensure that there is a solution to inequalities in Theorem 6.6;

Step 2: Based on Step 1, inequalities in Theorem 6.6 are in the form of LMIs, which can be solved by utilizing LMI Toolbox in Matlab. Then the control gain K_g is achieved.

Noting that the mentioned steps are also applicable to Theorem 6.9.

Remark 6.11 As mentioned in Remark 6.4, the studied extended dissipative performance is comprehensive, including the strict dissipative performance ($U_3 = R - \gamma I$ and $U_4 = 0$) and the $L_2 - L_\infty$ performance ($U_1 = 0,\ U_2 = 0,\ U_3 = \gamma^2 I$ and $U_4 = I$). Based on Step 1 in Remark 6.10, the dissipative and $L_2 - L_\infty$ controllers with the optimal performance index γ^* can be obtained by the following optimal approaches, respectively.
(1) Dissipative performance: for a given ε,

$$\begin{aligned} &\min \quad -\gamma \\ &\text{s.t.}\ \ \text{LMIs (6.26)} - \text{(6.29) in Theorem 6.6.} \end{aligned}$$

(2) $L_2 - L_\infty$ performance: for a given ε,

$$\begin{aligned} &\min \quad \gamma^2 \\ &\text{s.t.}\ \ \text{LMIs (6.26)} - \text{(6.29) in Theorem 6.6.} \end{aligned}$$

It is worth pointing out that the same designed algorithms also apply to Theorem 6.9.

Remark 6.12 In Theorems 6.6 and 6.9, the number of inequalities are $0.5(r + r^2)GN + N + rN$ and $0.5(r + r^2)GN + 1 + r$, respectively. We can observe that the computational cost of solving inequalities increases with the number of fuzzy rule r, switched mode N and observed mode G. It can be easily seen that the value r has much more influence on the calculation cost than the other two. On the other hand, the studied Marov process is traditional and its transition probabilities are constant and memoryless, which means that its process is not affected by the time spent in the mode. The limitation is conservative and hard to be satisfied in practice.

Compared with this, the semi-Markov process is more realistic because its transition probabilities are time-varying, dependent of the history/past information [13]. Thus, it is meaningful to investigate the asynchronous output feedback control problem for nonlinear semi-MJSs in future.

6.5 Illustrative Example

We still use the single-link robot arm system in Chap. 4 to illustrate the correctness of the presented design methods. The measurable output y_k, the output performance z_k and the external noise w_k are taken into consideration. The system is modelled as (6.1) with the following parameters:

$$
\begin{aligned}
&A_{1\delta_k} = \begin{bmatrix} 1 & T \\ -\frac{TglM_{\delta_k}}{J_{\delta_k}} & 1-\frac{TR}{J_{\delta_k}} \end{bmatrix},\ B_{11} = \begin{bmatrix} 0.1 \\ \frac{T}{J_1} \end{bmatrix},\ B_{21} = \begin{bmatrix} 0 \\ \frac{T}{J_1} \end{bmatrix},\\
&A_{2\delta_k} = \begin{bmatrix} 1 & T \\ -\frac{\beta TglM_{\delta_k}}{J_{\delta_k}} & 1-\frac{TR}{J_{\delta_k}} \end{bmatrix},\ B_{12} = B_{22} = \begin{bmatrix} 0 \\ \frac{T}{J_2} \end{bmatrix},\\
&C_{i\delta_k} = \begin{bmatrix} 0 \\ T \end{bmatrix},\ B_{13} = \begin{bmatrix} 0.5 \\ \frac{T}{J_3} \end{bmatrix},\ B_{23} = \begin{bmatrix} 0.5 \\ \frac{T}{J_3} \end{bmatrix},\ i = \{1, 2\},\\
&D_{i\delta_k} = \begin{bmatrix} 1 & 0 \end{bmatrix},\ E_{i\delta_k} = \begin{bmatrix} 1 & 0 \end{bmatrix},\ F_{i\delta_k} = 1,\ \delta_k = \{1, 2, 3\},\\
&M_1 = 1,\ M_2 = 5,\ M_3 = 10,\ g = 9.81,\ l = 0.5,\\
&J_1 = 1,\ J_2 = 5,\ J_3 = 10,\ R = 2,\\
&h_{1\delta_k}(x_{1k}) = \begin{cases} \dfrac{\sin(x_{1k}) - \beta x_{1k}}{(1-\beta)x_{1k}}, & x_{1k} \neq 0,\\ 1, & x_{1k} = 0, \end{cases}\\
&h_{2\delta_k}(x_{1k}) = 1 - h_{1\delta_k}(x_{1k}),\ \beta = \frac{0.01}{\pi},\ T = 0.1.
\end{aligned}
$$

The initial conditions are set as $x_0 = \begin{bmatrix} 0.5\pi & -0.1\pi \end{bmatrix}^T$ and $\delta_0 = 1$. The expectation of data dropouts is selected as $\alpha = 0.8$. The external disturbance is supposed to be

$$
w_k = \begin{cases} 0.5e^{0.1k}\sin(k), & 1 \le k \le 25,\\ 0, & \text{otherwise}. \end{cases}
$$

(1) For fuzzy MJSs, δ_k and η_k are assumed to obey the following probability matrices:

$$
\Pi = \begin{bmatrix} 0.3 & 0.2 & 0.5 \\ 0.4 & 0.2 & 0.4 \\ 0.55 & 0.15 & 0.3 \end{bmatrix},\ \Psi = \begin{bmatrix} 0.3 & 0.2 & 0.5 \\ 0.1 & 0.2 & 0.7 \\ 0.3 & 0.2 & 0.5 \end{bmatrix}.
$$

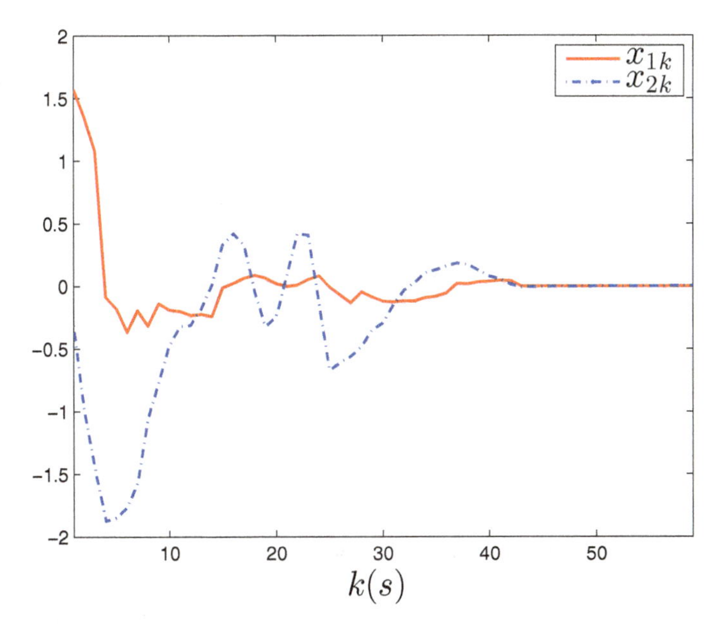

Fig. 6.1 State trajectories of the closed-loop system (6.7) with dissipative performance

Further, we assume that the dissipative performance indexes $U_1 = -1,\ U_2 = 1,\ U_3 = 1 - \gamma$ and $U_4 = 0$. By solving the inequalities with $\varepsilon = 0.1$ in Theorem 6.6, we have the following controller gains:

$$K_1 = -2.1118,\ K_2 = -1.8933,\ K_3 = -1.7929$$

with the optimal value $\gamma^* = 1.1327$. By assuming $U_1 = 0,\ U_2 = 0,\ U_3 = \gamma^2, U_4 = I$ and $F_{i\delta_k} = 0$, controller gains with $L_2 - L_\infty$ performance are computed as

$$K_1 = -2.2727,\ K_2 = -2.0759,\ K_3 = -1.9698$$

and the optimal value $\gamma^* = 0.1840$.

It is clearly observed from Fig. 6.1 that the devised control approach is effective to ensure that states converge to zero over time in the presence of the external disturbance and unreliable communication links. The similar results for the $L_2 - L_\infty$ control performance also hold, plotted in Fig. 6.2.

(2) For fuzzy switched systems with sojourn probabilities, we assume that $\bar{\beta}_1 = 0.3,\ \bar{\beta}_2 = 0.2,\ \bar{\beta}_3 = 0.5$ and the value of Ψ is unchanged. Based on Theorem 6.9 with $\varepsilon = 0.1, U_1 = -1,\ U_2 = 1,\ U_3 = 1 - \gamma$ and $U_4 = 0$, the dissipative controller gains are calculated as

$$K_1 = -2.4424,\ K_2 = -2.2178,\ K_3 = -2.1235$$

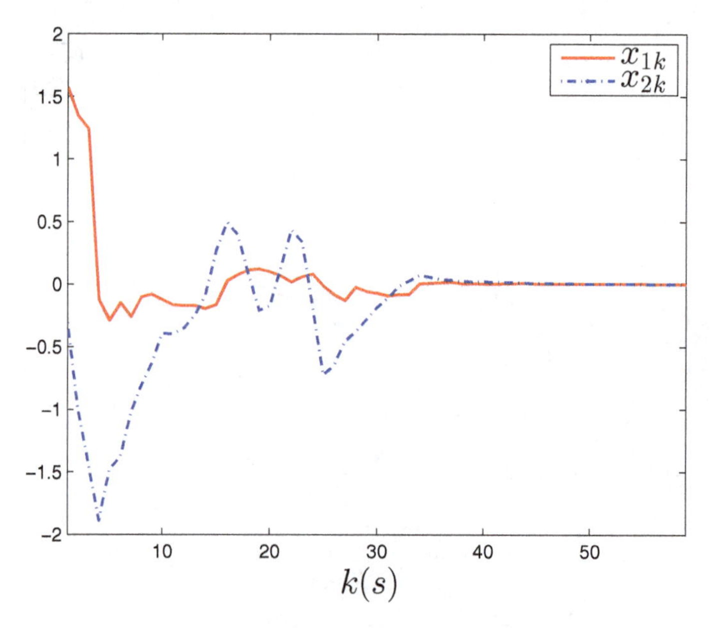

Fig. 6.2 State trajectories of the closed-loop system (6.7) with $L_2 - L_\infty$ performance

and the optimal dissipative performance $\gamma^* = 1.6996$. Further, let $U_1 = 0,\ U_2 = 0,\ U_3 = \gamma^2$, $U_4 = I$ and $F_{i\delta_k} = 0$. The $L_2 - L_\infty$ asynchronous controller gains are achieved as follows:

$$K_1 = -2.6403,\ K_2 = -2.6413,\ K_3 = -2.6410$$

and $\gamma^* = 0.1499$. The simulation results are plotted in Figs. 6.3 and 6.4, which demonstrate that our designed dissipative and $L_2 - L_\infty$ asynchronous output feedback controllers work well and our developed approach is correct and effective.

In the following, four different conditional transitional probability matrices Ψ^i $(i \in \{1, 2, 3, 4\})$ are assumed to show the influence of the synchronous, partially asynchronous and completely asynchronous phenomena on the system performance, including the dissipativity and $L_2 - L_\infty$ performance with

$$\Psi^1 = \begin{bmatrix} 1 & 0 & 0 \\ 0 & 1 & 0 \\ 0 & 0 & 1 \end{bmatrix},\ \Psi^2 = \begin{bmatrix} 1 & 0 & 0 \\ 0 & 1 & 0 \\ 0.1 & 0.2 & 0.7 \end{bmatrix},$$
$$\Psi^3 = \begin{bmatrix} 1 & 0 & 0 \\ 0.2 & 0.3 & 0.5 \\ 0.1 & 0.2 & 0.7 \end{bmatrix},\ \Psi^4 = \begin{bmatrix} 0.3 & 0.3 & 0.4 \\ 0.2 & 0.3 & 0.5 \\ 0.1 & 0.2 & 0.7 \end{bmatrix}.$$

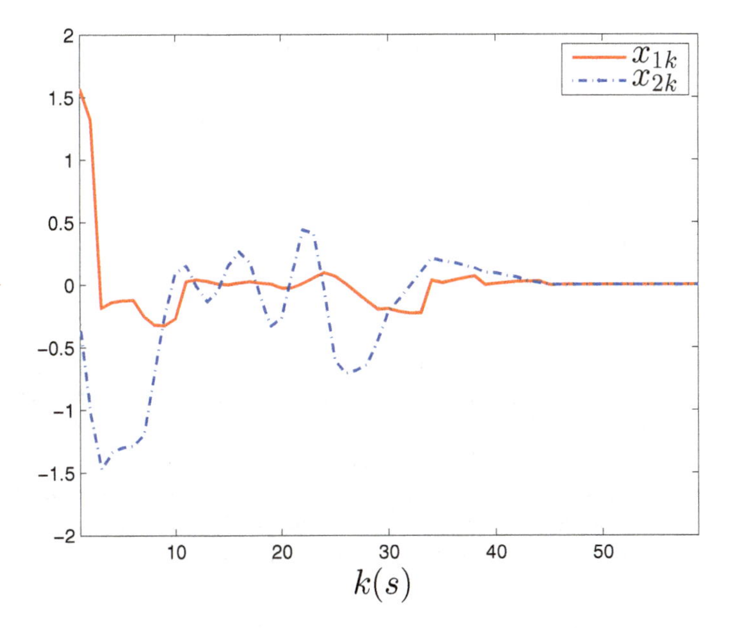

Fig. 6.3 State trajectories of the closed-loop system (6.34) with dissipative performance

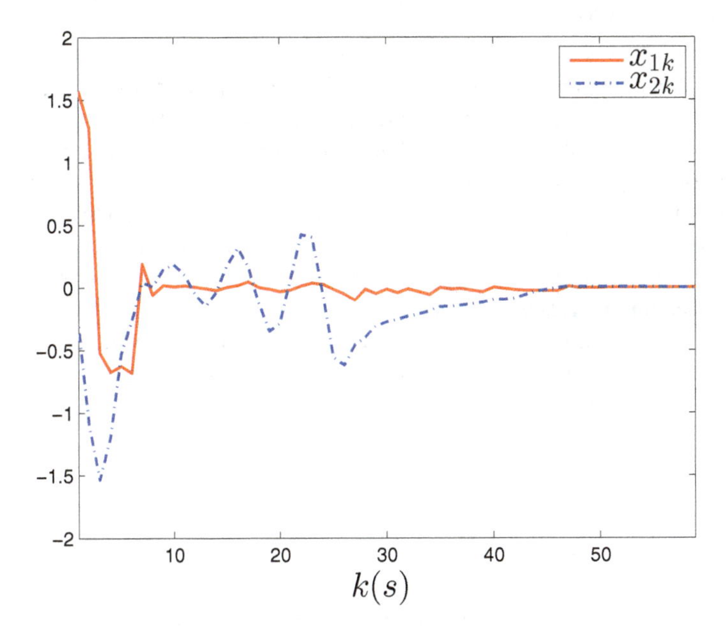

Fig. 6.4 State trajectories of the closed-loop system (6.34) with $L_2 - L_\infty$ performance

Table 6.1 Optimal γ^* with two performance indexes for both fuzzy switched systems (6.7) and (6.34) with $\varepsilon = 0.1$

γ^*	Ψ^1	Ψ^2	Ψ^3	Ψ^4
Dissipativity for (6.7)	1.6479	1.4202	1.3567	1.1790
$L_2 - L_\infty$ for (6.7)	0.1597	0.1717	0.1758	0.1824
Dissipativity for (6.34)	1.7440	1.7107	1.7030	1.7002
$L_2 - L_\infty$ for (6.34)	0.1455	0.1488	0.1494	0.1499

Recalling Remark 6.4, we assume that $U_1 = -1,\ U_2 = 1,\ U_3 = 1 - \gamma$ and $U_4 = 0$ for dissipative performance while $L_2 - L_\infty$ performance is considered with $U_1 = 0,\ U_2 = 0,\ U_3 = \gamma^2, U_4 = 1$ and $F_{i\delta_k} = 0$. From Table 6.1, we can find clearly that in the sense of dissipativity, both systems have the biggest γ^* when $\Psi = \Psi^1$. And as the increase of asynchronous levels, γ^* becomes smaller. There is an opposite trend for $L_2 - L_\infty$ performance. The simulation results imply that the synchronous case brings out the best behavior for both performance indexes.

6.6 Conclusion

In this chapter, we have analyzed the problem of the extended dissipative asynchronous output feedback control for fuzzy switched systems instead of using the state feedback control, where the HMM is employed to observe jump modes. The chapter covers two system models, namely, fuzzy MJSs with transition probabilities and another fuzzy switched system with sojourn probabilities. In addition, by using the notion of the extended dissipativity, the dissipative control (that contains passive control and H_∞ control) and $L_2 - L_\infty$ control can be designed in a unified framework. Based on the Lyapunov function, we have developed sufficient conditions to guarantee the stochastic stability of the resulting closed-loop systems with a satisfactory extended dissipative performance. The algorithms to solve controller gains are proposed as well.

References

1. Su, Q., Song, X.: Stabilization for a class of discrete-time switched systems with state constraints and quantized feedback. Int. J. Innov. Comput., Inf. Control **13**(6), 1829–1841 (2017)
2. Dolgov, M., Hanebeck, U.D.: Static output-feedback control of Markov jump linear systems without mode observation. IEEE Trans. Autom. Control **62**(10), 5401–5406 (2017)
3. Lin, H., Su, H., Chen, M.Z., Shu, Z., Lu, R., Wu, Z.-G.: On stability and convergence of optimal estimation for networked control systems with dual packet losses without acknowledgment. Automatica **90**, 81–90 (2018)

4. Lu, R., Wu, F., Xue, A.: Networked control with reset quantized state based on Bernoulli processing. IEEE Trans. Ind. Electron. **61**(9), 4838–4846 (2014)
5. Precup, R.-E., Tomescu, M.L., Preitl, S., Petriu, E.M., Fodor, J., Pozna, C.: Stability analysis and design of a class of MIMO fuzzy control systems. J. Intell. Fuzzy Syst. **25**(1), 145–155 (2013)
6. Radgolchin, M., Moeenfard, H.: Development of a multi-level adaptive fuzzy controller for beyond pull-in stabilization of electrostatically actuated microplates. J. Vib. Control **24**(5), 860–878 (2018)
7. Liu, Y., Park, J.H., Guo, B.-Z., Shu, Y.: Further results on stabilization of chaotic systems based on fuzzy memory sampled-data control. IEEE Trans. Fuzzy Syst. **26**(2), 1040–1045 (2018)
8. Chatterjee, A., Chatterjee, R., Matsuno, F., Endo, T.: Augmented stable fuzzy control for flexible robotic arm using LMI approach and neuro-fuzzy state space modeling. IEEE Trans. Ind. Electron. **55**(3), 1256–1270 (2008)
9. Vrkalovic, S., Teban, T.-A., Borlea, I.-D.: Stable Takagi-Sugeno fuzzy control designed by optimization. Int. J. Artif. Intell. **15**(2), 17–29 (2017)
10. Tanaka, K., Wang, H.O.: Fuzzy Control Systems Design and Analysis: a Linear Matrix Inequality Approach. Wiley, New Jersey (2004)
11. Wei, Y., Qiu, J., Lam, H.-K.: A novel approach to reliable output feedback control of fuzzy-affine systems with time delays and sensor faults. IEEE Trans. Fuzzy Syst. **25**(6), 1808–1823 (2017)
12. Wei, Y., Qiu, J., Shi, P., Wu, L.: A piecewise-Markovian Lyapunov approach to reliable output feedback control for fuzzy-affine systems with time-delays and actuator faults. IEEE Trans. Cybern. **48**(9), 2723–2735 (2018)
13. Wei, Y., Park, J.H., Qiu, J., Wu, L., Jung, H.Y.: Sliding mode control for semi-Markovian jump systems via output feedback. Automatica **81**, 133–141 (2017)
14. Zhou, J., Park, J.H., Kong, Q.: Robust resilient $L_2 - L_\infty$ control for uncertain stochastic systems with multiple time delays via dynamic output feedback. J. Frankl. Inst. **353**(13), 3078–3103 (2016)
15. Zhang, B., Zheng, W.X., Xu, S.: Filtering of Markovian jump delay systems based on a new performance index. IEEE Trans. Circuits Syst. I: Regul. Pap. **60**(5), 1250–1263 (2013)

Chapter 7
Dissipativity-Based Asynchronous Fuzzy Sliding Mode Control for Fuzzy MJSs

7.1 Introduction

As for control techniques, SMC is well-known as an effective robust control method in dealing with nonlinearities, uncertainties and noises. Its purpose is to devise an applicable control scheme to ensure that system trajectories are driven onto the user-defined sliding surface in limited time and subsequently operate there. Thus, it has been extensively used, for instance, the finite-time stabilization for continuous-time nonlinear systems [1] and the event-triggered control for stochastic systems [2]. On the other hand, in approximating nonlinear systems by adopting the T–S fuzzy model, some model uncertainties or errors are created. Due to the quick reaction and powerful robustness of SMC, it has been developed to cope with model uncertainties or errors in T–S fuzzy systems. In [3], a classic SMC and a non-PDC SMC scheme have been devised to investigate the robust stabilization issue for T–S fuzzy stochastic descriptor systems. The work in [4] has addressed the robust SMC issue for T–S fuzzy systems with mismatched and matched uncertainties. For T–S fuzzy systems with uncertainties and noises, a new fuzzy integral SMC surface has been constructed to deal with the dissipativity-based control problem in [5].

In this chapter, the dissipative-based control problem for T–S fuzzy systems with matched uncertainties and unavailability of jump modes is investigated by using the AFISMC approach. With the observed mode by the HMM, a novel fuzzy integral sliding surface is designed with the PDC approach. We employ the mode-dependent Lyapunov function to analyze the stochastic stability of the sliding mode dynamics with a given dissipative performance and develop a solution to controller gains simultaneously. An AFISMC law and an adaptive AFISMC law are constructed to ensure the operation of system trajectories toward a small bounded region, respectively. A numerical example is given to verify the correctness and validity of the developed AFISMC techniques.

S. Dong et al., *Control and Filtering of Fuzzy Systems with Switched Parameters*, Studies in Systems, Decision and Control 268,
https://doi.org/10.1007/978-3-030-35566-1_7

7.2 Preliminary Analysis

Considering the following continuous-time T–S fuzzy MJSs on the probability space $(\Omega, \mathcal{F}, \mathcal{P})$:
Plant rule $i^{\alpha(t)}$: IF $\rho_1(t)$ is $\varphi_{i1}^{\alpha(t)}$, and ..., and $\rho_q(t)$ is $\varphi_{iq}^{\alpha(t)}$, THEN

$$\begin{cases} \dot{x}(t) = A_{i\alpha(t)}x(t) + B_{i\alpha(t)}(u(t) + f(x(t))) + C_{i\alpha(t)}w(t), \\ z(t) = D_{i\alpha(t)}x(t) + E_{i\alpha(t)}w(t), \end{cases} \tag{7.1}$$

where $x(t) \in R^{n_x}$, $u(t) \in R^{n_u}$, $w(t) \in R^{n_w}$ and $z(t) \in R^{n_z}$ are the state, the control input, the exterior noise belonging to $l_2[0, +\infty)$ and the output, respectively. i ($i \in \mathcal{R} = \{1, \ldots, r\}$) is the ith fuzzy rule. $\rho_1(t), \ldots, \rho_q(t)$ are the premise variables and $\varphi_{i1}^{\alpha(t)}, \ldots, \varphi_{iq}^{\alpha(t)}$ are the fuzzy sets. The nonlinear term $f(x(t))$ denotes the matched uncertainty with $\|f(x(t))\| < \kappa\|x(t)\|$, where κ is given. The stochastic variable $\alpha(t)$ ($\alpha(t) \in \mathcal{M}_1 = \{1, \ldots, M_1\}$) is used to represent the continuous-time Markov process with right continuous trajectories, subject to the transition probability matrix $\Theta = [\theta_{ab}]$ and

$$\Pr\{\alpha(t+dt) = b|\alpha(t) = a\} = \begin{cases} \theta_{ab}dt + o(dt), \quad a \neq b, \\ 1 + \theta_{aa}dt + o(dt), a = b, \end{cases} \tag{7.2}$$

where $\theta_{ab} \geq 0$ if $a \neq b$ and $\theta_{bb} = -\sum_{a=1,a\neq b}^{M_1} \theta_{ab}$. dt denotes the infinitesimal transition time period with $\lim_{dt\to 0} \frac{o(dt)}{dt} = 0$.

By letting $\alpha(t) = a$, the overall fuzzy system from (7.1) is inferred as

$$\begin{cases} \dot{x}(t) = A_{ha}(t)x(t) + B_{ha}(t)(u(t) + f(x(t))) + C_{ha}(t)w(t), \\ z(t) = D_{ha}(t)x(t) + E_{ha}(t)w(t), \end{cases} \tag{7.3}$$

where

$$A_{ha}(t) = \sum_{i=1}^{r} h_{ia}(t)A_{ia}, \ B_{ha}(t) = \sum_{i=1}^{r} h_{ia}(t)B_{ia},$$

$$C_{ha}(t) = \sum_{i=1}^{r} h_{ia}(t)C_{ia}, \ D_{ha}(t) = \sum_{i=1}^{r} h_{ia}(t)D_{ia},$$

$$E_{ha}(t) = \sum_{i=1}^{r} h_{ia}(t)E_{ia}, \ \sum_{i=1}^{r} h_{ia}(t) = 1,$$

$$h_{ia}(t) = \frac{\prod_{j=1}^{q} \varphi_{ij}^{a}(\rho_j(t))}{\sum_{i=1}^{r}\prod_{j=1}^{q} \varphi_{ij}^{a}(\rho_j(t))} \geq 0.$$

$h_{ia}(t)$ is the normalized membership function and $\varphi_{ij}^{a}(\rho_j(t))$ describes the grade of membership of premise variable $\rho_j(t)$ in φ_{ij}^{a}. It is assumed that $B_{ha}(t)$ has full column rank, that is, $rank(B_{ha}(t)) = n_u$.

The main objective of this chapter is to construct a novel integral SMC approach such that the sliding mode dynamics of (7.3) is stochastically stable with a desired dissipative performance.

7.3 Sliding Mode Control

In this section, we firstly design a new continuous-time AFISMC surface, which can preferably adapt to the features of T–S fuzzy MJSs. Then, we focus on designing an approach to solve the sliding mode controller gains, which can guarantee the desired performance of the corresponding sliding mode dynamics. By using the obtained controller gains, AFISMC laws are presented to ensure the reachability of system trajectories in (7.3) toward the bounded region of the predetermined sliding surface in finite time.

7.3.1 Sliding Surface Design

Since $\alpha(t)$ cannot be accessed easily, the HMM is adopted to observe $\alpha(t)$ with the following model:

$$\Pr\{\beta(t) = c|\alpha(t) = a\} = \varrho_{ac}, \tag{7.4}$$

where $\beta(t)$ takes a value in $\mathcal{M}_2 = \{1, \ldots, M_2\}$ and satisfies the conditional transition probability matrix $\Gamma = [\varrho_{ac}]$ with $\varrho_{ac} \geq 0$ and $\sum_{c=1}^{M_2} \varrho_{ac} = 1$.

Then using the estimated mode $\beta(t)$ and the PDC approach, we construct the following continuous-time AFISMC surface:

$$\begin{aligned} s(t) = {} & B_{ha}^T(0)x(0) + \int_0^t B_{ha}^T(v)\dot{x}(v)dv \\ & - \int_0^t B_{ha}^T(v)(A_{ha}(v) + B_{ha}(v)K_{h\beta(v)}(v))x(v)dv, \end{aligned} \tag{7.5}$$

where $K_{h\beta(t)}(t)$ is the sliding mode controller to be designed. According to the PDC principle, $K_{h\beta(t)}(t)$ has the same fuzzy rules as (7.3) and it further derives that $K_{h\beta(t)}(t) = \sum_{i=1}^{r} h_{ia}(t)K_{i\beta(t)}$. Taking the derivative of $s(t)$ with $\beta(t) = c$, we have

$$\begin{aligned} \dot{s}(t) = {} & B_{ha}^T(t)B_{ha}(t)(u(t) + f(x(t))) + B_{ha}^T(t)C_{ha}(t)w(t) \\ & - B_{ha}^T(t)B_{ha}(t)K_{hc}(t)x(t). \end{aligned} \tag{7.6}$$

Remark 7.1 In designing a suitable sliding surface for T–S fuzzy MJSs, we take the unavailability of jump modes into account and use the HMM principle to estimate $\alpha(t)$. With the estimated $\beta(t)$, the new asynchronous fuzzy sliding surface is constructed as (7.5) and an HMM $(\alpha(t), \beta(t), \Theta, \Gamma)$ is built. The so-called asynchronous situation is that jump modes between the controller and original systems are different, namely, $\alpha(t) \neq \beta(t)$. And the synchronous controller $(\alpha(t) = \beta(t))$ and the mode independent controller $(\beta(t) \in \{1\})$ can be seen as two special cases of non-synchronization.

Assume that the predefined surface is $s(t) = 0$. It easily follows that $\dot{s}(t) = 0$. Hence, the equivalent control input is obtained as

$$u_{eq}(t) = K_{hc}(t)x(t) - f(x(t)) - (B_{ha}^T(t)B_{ha}(t))^{-1}B_{ha}^T(t)C_{ha}(t)w(t). \tag{7.7}$$

It should be mentioned that due to $rank(B_{ha}(t)) = n_u$, we have that $B_{ha}^T(t)B_{ha}(t)$ is a nonsingular and positive-definite matrix.

Using the input $u_{eq}(t)$, we have the following sliding mode dynamics:

$$\begin{cases} \dot{x}(t) = \hat{A}_{hac}(t)x(t) + \tilde{B}_{ha}(t)C_{ha}(t)w(t), \\ z(t) = D_{ha}(t)x(t) + E_{ha}(t)w(t), \end{cases} \tag{7.8}$$

where

$$\hat{A}_{hac}(t) = A_{ha}(t) + B_{ha}(t)K_{hc}(t),$$
$$\tilde{B}_{ha}(t) = I - B_{ha}(t)(B_{ha}^T(t)B_{ha}(t))^{-1}B_{ha}^T(t).$$

7.3.2 Analysis on Sliding Mode Dynamics

In the following, we firstly present a sufficient condition for the desired performance of system (7.8) and then develop a solution to the sliding mode controller gain K_{jc}.

Theorem 7.1 *System (7.8) is stochastically stable with a desired dissipative performance* (G_1, G_2, G_3, δ) *if there exist matrices* $P_a > 0$, K_{jc}, *scalar* $\varepsilon > 0$ *for all* $i, j \in \mathcal{R}$, $a \in \mathcal{M}_1$ *and* $c \in \mathcal{M}_2$ *satisfying*

$$\Omega_{iia} < 0, \tag{7.9}$$

$$\Omega_{ija} + \Omega_{jia} < 0, \; i < j, \tag{7.10}$$

where

$$\Omega_{ija} = \begin{bmatrix} \Omega_{ija}^{11} & * & * & * & * \\ -G_2^T D_{ia} & \Omega_{ia}^{22} & * & * & * \\ G_1^+ D_{ia} & G_1^+ E_{ia} & -I & * & * \\ P_a & 0 & 0 & -\varepsilon I & * \\ 0 & \varepsilon C_{ia} & 0 & 0 & -\varepsilon I \end{bmatrix},$$

$$\begin{aligned} \Omega_{ija}^{11} &= P_a(A_{ia} + \sum_{c=1}^{M_2} \varrho_{ac} B_{ia} K_{jc}) \\ &\quad + (A_{ia} + \sum_{c=1}^{M_2} \varrho_{ac} B_{ia} K_{jc})^T P_a + \sum_{b=1}^{M_1} \theta_{ab} P_b, \\ \Omega_{ia}^{22} &= -E_{ia}^T G_2 - G_2^T E_{ia} - G_3 + \delta I. \end{aligned}$$

Proof Due to $h_{ia}(t) \geq 0$ and $\sum_{i=1}^{r} h_{ia}(t) = 1$, we have

$$\begin{aligned} \Omega_{ha}(t) &= \sum_{i=1}^{r} \sum_{j=1}^{r} h_{ia}(t) h_{ja}(t) \Omega_{ija} \\ &= \sum_{i=1}^{r} h_{ia}^2(t) \Omega_{iia} + \sum_{i=1}^{r-1} \sum_{j=i+1}^{r} h_{ia}(t) h_{ja}(t) (\Omega_{ija} + \Omega_{jia}) \\ &< 0, \end{aligned} \tag{7.11}$$

where

$$\Omega_{ha}(t) = \begin{bmatrix} \Omega_{ha}^{11} & * & * & * & * \\ -G_2^T D_{ha}(t) & \Omega_{ha}^{22} & * & * & * \\ G_1^+ D_{ha}(t) & G_1^+ E_{ha}(t) & -I & * & * \\ P_a & 0 & 0 & -\varepsilon I & * \\ 0 & \varepsilon C_{ha}(t) & 0 & 0 & -\varepsilon I \end{bmatrix},$$

$$\begin{aligned} \Omega_{ha}^{11} &= \sum_{c=1}^{M_2} \varrho_{ac} (P_a \hat{A}_{hac}(t) + \hat{A}_{hac}^T(t) P_a) + \sum_{b=1}^{M_1} \theta_{ab} P_b, \\ \Omega_{ha}^{22} &= -E_{ha}^T G_2 - G_2^T E_{ha} - G_3 + \delta I. \end{aligned}$$

By using Schur Complement to the above inequality, it follows that

$$\begin{aligned} \Omega_1(t) &= \begin{bmatrix} \Omega_{ha}^{11} & * \\ -G_2^T D_{ha}(t) & \Omega_{ha}^{22} \end{bmatrix} - \begin{bmatrix} D_{ha}^T(t) \\ E_{ha}^T(t) \end{bmatrix} G_1 \begin{bmatrix} D_{ha}^T(t) \\ E_{ha}^T(t) \end{bmatrix}^T \\ &\quad + \varepsilon^{-1} \begin{bmatrix} P_a \\ 0 \end{bmatrix} \begin{bmatrix} P_a \\ 0 \end{bmatrix}^T + \varepsilon \begin{bmatrix} 0 \\ C_{ha}^T(t) \end{bmatrix} \begin{bmatrix} 0 \\ C_{ha}^T(t) \end{bmatrix}^T < 0. \end{aligned} \tag{7.12}$$

Due to the fact that $\tilde{B}_{ha}(t) = \tilde{B}_{ha}^T(t) = I - B_{ha}(t)(B_{ha}^T(t)B_{ha}(t))^{-1}B_{ha}^T(t)$, $\tilde{B}_{ha}^T(t)$ $\tilde{B}_{ha}(t) = \tilde{B}_{ha}(t)$ and the nonsingular property of $B_{ha}^T(t)B_{ha}(t)$, we can clearly conclude that $\tilde{B}_{ha}^T(t)\tilde{B}_{ha}(t) \leq I$. By employing Lemma 1 in [6], the equivalent relationship between $\Omega_1(t) < 0$ and $\Omega_2(t) < 0$ is established with

$$\Omega_2(t) = \begin{bmatrix} \Omega_{ha}^{11} & * \\ -G_2^T D_{ha}(t) & \Omega_{ha}^{22} \end{bmatrix} - \begin{bmatrix} D_{ha}^T(t) \\ E_{ha}^T(t) \end{bmatrix} G_1 \begin{bmatrix} D_{ha}^T(t) \\ E_{ha}^T(t) \end{bmatrix}^T$$
$$+ sym(\begin{bmatrix} P_a \\ 0 \end{bmatrix} \tilde{B}_{ha}(t) \begin{bmatrix} 0 \\ C_{ha}^T(t) \end{bmatrix}^T) < 0,$$

where $sym(X) = X + X^T$.

On the other hand, Lyapunov function is chosen as

$$V(t) = x^T(t)P_{\alpha(t)}x(t) \tag{7.13}$$

where $P_{\alpha(t)} > 0$. With $w(t) \equiv 0$ along trajectories of system (7.8), it follows that

$$\mathcal{D}V(t) = x^T(t)\Omega_{ha}^{11}x(t) \tag{7.14}$$

where $\mathcal{D}$ is the weak infinitesimal generator of the stochastic process. It follows from (7.11) that $\Omega_{ha}^{11} < 0$, which means that the stochastic stability of system (7.8) is ensured. When $w(t) \neq 0$, we have

$$\begin{aligned} &E\{\mathcal{D}V(\upsilon) - J(\upsilon) + \delta w^T(\upsilon)w(\upsilon)\} \\ &= \begin{bmatrix} x(\upsilon) \\ w(\upsilon) \end{bmatrix}^T \Omega_2(t) \begin{bmatrix} x(\upsilon) \\ w(\upsilon) \end{bmatrix} < 0. \end{aligned} \tag{7.15}$$

Under the zero initial condition, we achieve that

$$\int_0^t E\{J(\upsilon)\}d\upsilon > \delta \int_0^t E\{w^T(\upsilon)w(\upsilon)\}d\upsilon + E\{V(t)\}.$$

Thus, the dissipative performance is obtained. The proof is completed.

It is clearly observed from Theorem 7.1 that there exists the nonlinear coupling between P_a and K_{jc}, which leads to computational difficulty when computing K_{jc}. The following theorem presents a easier solution to K_{jc}.

Theorem 7.2 *System (7.8) is stochastically stable with a desired dissipative performance* (G_1, G_2, G_3, δ) *if there exist matrices* $\bar{P}_a > 0$, X, Y_{jc}, *scalars* $\varepsilon > 0$ *and* $\gamma > 0$ *for all* i, $j \in \mathcal{R}$, $a \in \mathcal{M}_1$ *and* $c \in \mathcal{M}_2$ *satisfying*

$$\Xi_{iia} < 0, \tag{7.16}$$

$$\Xi_{ija} + \Xi_{jia} < 0, \ i < j, \tag{7.17}$$

where

$$\Xi_{ija} = \begin{bmatrix} \Xi^{11} & * & * & * & * & * & * & * \\ \Xi_{ija}^{21} & \Xi_a^{22} & * & * & * & * & * & * \\ \Xi_{ia}^{31} & 0 & \Xi_{ia}^{33} & * & * & * & * & * \\ \Xi_{ia}^{41} & 0 & \Xi_{ia}^{43} & -I & * & * & * & * \\ 0 & I & 0 & 0 & -\varepsilon I & * & * & * \\ 0 & 0 & \varepsilon C_{ia} & 0 & 0 & -\varepsilon I & * & * \\ X & 0 & 0 & 0 & 0 & 0 & -\gamma \bar{P}_a & * \\ 0 & \Xi_a^{82} & 0 & 0 & 0 & 0 & 0 & \Xi^{88} \end{bmatrix},$$

$$\Xi^{11} = -X - X^T, \; \Xi_{ija}^{21} = A_{ia}X + \sum_{c=1}^{M_2} \varrho_{ac} B_{ia} Y_{jc} + \bar{P}_a,$$
$$\Xi_a^{22} = -\gamma^{-1}\bar{P}_a + \theta_{aa}\bar{P}_a, \; \Xi_{ia}^{31} = -G_2^T D_{ia} X,$$
$$\Xi_{ia}^{33} = -E_{ia}^T G_2 - G_2^T E_{ia} - G_3 + \delta I,$$
$$\Xi_{ia}^{41} = G_1^+ D_{ia} X, \; \Xi_{ia}^{43} = G_1^+ E_{ia},$$
$$\Xi_a^{82} = [\theta_{a1}^{0.5}\bar{P}_a, \ldots, \theta_{aa-1}^{0.5}\bar{P}_a, \theta_{aa+1}^{0.5}\bar{P}_a, \ldots, \theta_{aM_1}^{0.5}\bar{P}_a]^T,$$
$$\Xi^{88} = -diag\{\bar{P}_1, \ldots, \bar{P}_{a-1}, \bar{P}_{a+1}, \ldots, \bar{P}_{M_1}\}.$$

Moreover, the controller gain can be computed from

$$K_{jc} = Y_{jc}X^{-1}. \tag{7.18}$$

Proof From (7.16), (7.17), it follows that

$$\begin{aligned} \Xi_{ha}(t) &= \sum_{i=1}^{r}\sum_{j=1}^{r} h_{ia}(t)h_{ja}(t)\Xi_{ija} \\ &= \sum_{i=1}^{r} h_{ia}^2(t)\Xi_{iia} + \sum_{i=1}^{r-1}\sum_{j=i+1}^{r} h_{ia}(t)h_{ja}(t)(\Xi_{ija} + \Xi_{jia}) \\ &< 0, \end{aligned} \tag{7.19}$$

where

$$\Xi_{ha}(t) = \begin{bmatrix} \Xi^{11} & * & * & * & * & * & * & * \\ \Xi_{ha}^{21} & \Xi_a^{22} & * & * & * & * & * & * \\ \Xi_{ha}^{31} & 0 & \Xi_{ha}^{33} & * & * & * & * & * \\ \Xi_{ha}^{41} & 0 & \Xi_{ha}^{43} & -I & * & * & * & * \\ 0 & I & 0 & 0 & -\varepsilon I & * & * & * \\ 0 & 0 & \varepsilon C_{ha}(t) & 0 & 0 & -\varepsilon I & * & * \\ X & 0 & 0 & 0 & 0 & 0 & -\gamma \bar{P}_a & * \\ 0 & \Xi_a^{82} & 0 & 0 & 0 & 0 & 0 & \Xi^{88} \end{bmatrix},$$

$$\Xi_{ha}^{21} = A_{ha}(t)X + \sum_{c=1}^{M_2} \varrho_{ac} B_{ha}(t)Y_{hc}(t) + \bar{P}_a,$$
$$\Xi_{ha}^{31} = -G_2^T D_{ha}(t)X,$$
$$\Xi_{ha}^{33} = -E_{ha}^T(t)G_2 - G_2^T E_{ha}(t) - G_3 + \delta I,$$
$$\Xi_{ha}^{41} = G_1^+ D_{ha}(t)X, \ \Xi_{ha}^{43} = G_1^+ E_{ha}(t).$$

Using Schur Complement to (7.19), we have

$$\Xi_{ha}^1(t) < 0, \tag{7.20}$$

where

$$\Xi_{ha}^1(t) = \begin{bmatrix} \Xi^{11} & * & * & * & * & * & * \\ \Xi_{ha}^{21} & \bar{\Xi}_a^{22} & * & * & * & * & * \\ \Xi_{ha}^{31} & 0 & \Xi_{ha}^{33} & * & * & * & * \\ \Xi_{ha}^{41} & 0 & \Xi_{ha}^{43} & -I & * & * & * \\ 0 & I & 0 & 0 & -\varepsilon I & * & * \\ 0 & 0 & \varepsilon C_{ha}(t) & 0 & 0 & -\varepsilon I & * \\ X & 0 & 0 & 0 & 0 & 0 & -\gamma \bar{P}_a \end{bmatrix},$$
$$\bar{\Xi}_a^{22} = -\gamma^{-1}\bar{P}_a + \sum_{b=1}^{M_1} \theta_{ab} \bar{P}_a \bar{P}_b^{-1} \bar{P}_a.$$

Define

$$\mathcal{J}(t) = \begin{bmatrix} \tilde{A}_{hac}(t) & I\,0\,0\,0\,0\,0 \\ -G_2^T D_{ha}(t) & 0\,I\,0\,0\,0\,0 \\ G_1^+ D_{ha}(t) & 0\,0\,I\,0\,0\,0 \\ 0 & 0\,0\,0\,I\,0\,0 \\ 0 & 0\,0\,0\,0\,I\,0 \\ I & 0\,0\,0\,0\,0\,I \end{bmatrix}, \tag{7.21}$$
$$\mathcal{I} = \mathrm{diag}\{P_a, I, I, I, I, I\}, \ P_a = \bar{P}_a^{-1},$$

with $\tilde{A}_{hac}(t) = A_{ha}(t) + \sum_{c=1}^{M_2} \varrho_{ac} B_{ha}(t) K_{hc}(t)$. Due to $K_{jc} = Y_{jc} X^{-1}$, we derive that

$$\mathcal{I}\mathcal{J}(t)\Xi_{hac}^1 \mathcal{J}^T(t)\mathcal{I} = \Omega^1(t) < 0, \tag{7.22}$$

where

$$\Omega^1(t) = \begin{bmatrix} \Omega_{ha}^{11} - \gamma^{-1}P_a & * & * & * & * & * \\ -G_2^T D_{ha}(t) & \Xi_{ha}^{33} & * & * & * & * \\ G_1^+ D_{ha}(t) & G_1^+ E_{ha}(t) & -I & * & * & * \\ P_a & 0 & 0 & -\varepsilon I & * & * \\ 0 & \varepsilon C_{ha}(t) & 0 & 0 & -\varepsilon I & * \\ I & 0 & 0 & 0 & 0 & -\gamma P_a^{-1} \end{bmatrix}.$$

By using Schur Complement again to the above inequality, it follows that (7.12) holds. Thus, it is clearly concluded that the inequalities in Theorem 7.2 can ensure the correctness of those in Theorem 7.1. Accordingly, if there exists a feasible solution to inequalities in Theorem 7.2, the controller gains can be computed from $K_{jc} = Y_{jc}X^{-1}$.

7.3.3 Sliding Mode Control Law

By employing K_{ic} in Theorem 7.1 (or Theorem 7.2), an AFISMC law is presented to guarantee the reachability of trajectories in (7.3) to the pre-determined surface $s(t) = 0$ in finite time.

Theorem 7.3 *The trajectories in (7.3) can be forced onto the specified surface $s(t) = 0$ in finite time via the following AFISMC law:*

$$u(t) = K_{hc}(t)x(t) - \phi(t)sign(\hat{B}_{ha}(t)s(t)), \tag{7.23}$$

with

$$\begin{aligned} &K_{hc}(t) = \sum_{i=1}^{r} h_{ia}(t)K_{ic}, \ \hat{B}_{ha}(t) = B_{ha}^T(t)B_{ha}(t), \ \phi_1 > 0, \\ &\phi(t) = (\phi_1 + \kappa\phi_2 + \phi_3)/\phi_4, \ \phi_2 = \|\hat{B}_{ha}(t)\| \|x(t)\|, \\ &\phi_3 = \|B_{ha}^T(t)C_{ha}(t)w(t)\|, \ \phi_4 = \sqrt{\lambda_{min}(\hat{B}_{ha}(t)\hat{B}_{ha}(t))}. \end{aligned}$$

Proof The candidate Lyapunov function is

$$V(t) = 0.5s^T(t)s(t). \tag{7.24}$$

Based on the surface (7.6) and the control law (7.23), we have

$$\begin{aligned} \dot{s}(t) = & -\phi(t)\hat{B}_{ha}(t)sign(\hat{B}_{ha}(t)s(t)) + \hat{B}_{ha}(t)f(x(t)) \\ & + B_{ha}^T(t)C_{ha}(t)w(t). \end{aligned} \tag{7.25}$$

Then, based on the fact of $\|\hat{B}_{ha}(t)s(t)\| < s^T(t)\hat{B}_{ha}^T(t)sign(\hat{B}_{ha}(t)s(t))$, it derives that

$$
\begin{aligned}
E\{\dot{V}(t)\} &= E\{s^T(t)\dot{s}(t)\} \\
&= -\phi(t)s^T(t)\hat{B}_{ha}(t)sign(\hat{B}_{ha}(t)s(t)) \\
&\quad + s^T(t)\hat{B}_{ha}(t)f(x(t)) + s^T(t)B_{ha}^T(t)C_{ha}(t)w(t) \\
&< -\phi(t)\|\hat{B}_{ha}(t)s(t)\| + \kappa\phi_2\|s(t)\| + \phi_3\|s(t)\| \\
&< -\phi(t)\phi_4\|s(t)\| + \kappa\phi_2\|s(t)\| + \phi_3\|s(t)\| \\
&< -\phi_1\|s(t)\|.
\end{aligned} \tag{7.26}
$$

We can see that trajectories in (7.3) can be forced onto the predetermined surface $s(t) = 0$ in finite time. This completes the proof.

A disadvantage of the AFISMC law (7.23) is the discontinuous feature at surface $s(t) = 0$, which can induce the unsatisfactory chattering issue. It is best to make the control law smooth in practical operation. For smoothing the control law, a boundary layer is introduced to design the following control strategy [7].

Theorem 7.4 *The trajectories in (7.3) can be forced onto a small bounded region around the specified surface $s(t) = 0$ in finite time via the following AFISMC law:*

$$
u(t) = K_{hc}(t)x(t) - \mu s(t) + \chi(t) \cdot o(s(t)), \tag{7.27}
$$

with

$$
\begin{aligned}
o(s(t)) &= \begin{cases} -\dfrac{\hat{B}_{ha}(t)s(t)}{\|\hat{B}_{ha}(t)s(t)\|}, & \|\hat{B}_{ha}(t)s(t)\| > o, \\ -\dfrac{\hat{B}_{ha}(t)s(t)}{o}, & \|\hat{B}_{ha}(t)s(t)\| \leq o, \end{cases} \\
K_{hc}(t) &= \sum_{i=1}^{r} h_{ia}(t)K_{ic}, \ \hat{B}_{ha}(t) = B_{ha}^T(t)B_{ha}(t), \ \mu > 0, \\
\chi(t) &= (\kappa\chi_1 + \chi_2)/\chi_3, \ \chi_1 = \|\hat{B}_{ha}(t)\|\|x(t)\|, \\
\chi_2 &= \|B_{ha}^T(t)C_{ha}(t)w(t)\|, \ \chi_3 = \sqrt{\lambda_{min}(\hat{B}_{ha}(t)\hat{B}_{ha}(t))}.
\end{aligned}
$$

Proof The candidate Lyapunov function is

$$
V(t) = 0.5s^T(t)s(t). \tag{7.28}
$$

It easily follows from (7.6) and (7.27) that

$$
\begin{aligned}
\dot{s}(t) = &- \mu\hat{B}_{ha}(t)s(t) + \chi(t)\hat{B}_{ha}(t) \cdot o(s(t)) \\
&+ \hat{B}_{ha}(t)f(x(t)) + B_{ha}^T(t)C_{ha}(t)w(t).
\end{aligned} \tag{7.29}
$$

When $\|\hat{B}_{ha}(t)s(t)\| > o$,

$$\begin{aligned} E\{\dot{V}(t)\} &= E\{s^T(t)\dot{s}(t)\} \\ &< -\mu s^T(t)\hat{B}_{ha}(t)s(t) - \chi(t)\|\hat{B}_{ha}(t)s(t))\| + (\kappa\chi_1 + \chi_2)\|s(t)\| \\ &< -\mu s^T(t)\hat{B}_{ha}(t)s(t) - \chi(t)\chi_3\|s(t))\| + (\kappa\chi_1 + \chi_2)\|s(t)\| \\ &< -\mu s^T(t)\hat{B}_{ha}(t)s(t). \end{aligned} \tag{7.30}$$

When $\|\hat{B}_{ha}(t)s(t)\| \leq o$,

$$\begin{aligned} E\{\dot{V}(t)\} &= E\{s^T(t)\dot{s}(t)\} \\ &< -\mu s^T(t)\hat{B}_{ha}(t)s(t) - \frac{\chi(t)}{o}\|\hat{B}_{ha}(t)s(t))\|^2 + (\kappa\chi_1 + \chi_2)\|s(t)\| \\ &< -\mu s^T(t)\hat{B}_{ha}(t)s(t) + (\kappa\chi_1 + \chi_2)\|s(t)\|(1 - \frac{\chi_3\|s(t))\|}{o}). \end{aligned} \tag{7.31}$$

We can choose a suitable value μ to guarantee that $E\{\dot{V}(t)\} < 0$ and to make that $s(t)$ runs in a bounded region including $s(t) = 0$. Thus, state trajectories can work in the sliding mode region around the specified surface $s(t) = 0$ in finite time through the AFISMC law (7.27).

The upper bound κ of the uncertainty term $f(x(t))$ may be unknown in practice. An adaptive AFISMC law is given to deal with this problem, as follows:

Theorem 7.5 *The trajectories in (7.3) can be forced onto a small bounded region around the specified surface $s(t) = 0$ in finite time via the following AFISMC law:*

$$u(t) = K_{hc}(t)x(t) - \mu s(t) + \hat{\chi}(t) \cdot o(s(t)), \tag{7.32}$$

with

$$o(s(t)) = \begin{cases} -\dfrac{\hat{B}_{ha}(t)s(t)}{\|\hat{B}_{ha}(t)s(t)\|}, & \|\hat{B}_{ha}(t)s(t)\| > o, \\ -\dfrac{\hat{B}_{ha}(t)s(t)}{o}, & \|\hat{B}_{ha}(t)s(t)\| \leq o, \end{cases}$$

$$K_{hc}(t) = \sum_{i=1}^{r} h_{ia}(t)K_{ic}, \;\; \hat{B}_{ha}(t) = B_{ha}^T(t)B_{ha}(t), \;\; \mu > 0,$$

$$\hat{\chi}(t) = (\hat{\kappa}(t)\chi_1 + \chi_2)/\chi_3, \;\; \dot{\hat{\kappa}}(t) = \frac{1}{\varpi}\chi_1\|s(t)\|, \;\; \varpi > 0,$$

$$\chi_1 = \|\hat{B}_{ha}(t)\|\|x(t)\|, \;\; \chi_2 = \|B_{ha}^T(t)C_{ha}(t)w(t)\|,$$

$$\chi_3 = \sqrt{\lambda_{min}(\hat{B}_{ha}(t)\hat{B}_{ha}(t))}.$$

Proof Define Lyapunov function as

$$V(t) = 0.5s^T(t)s(t) + 0.5\varpi\tilde{\kappa}^2(t) \tag{7.33}$$

with $\tilde{\kappa}(t) = \hat{\kappa}(t) - \kappa$. By similar proof process in Theorem 7.4, we obtain Theorem 7.5. Hence, the proof is omitted.

7.4 Extension to Discrete-Time Fuzzy MJSs

Recently, most published works mainly focus on the fuzzy integral sliding surface design for continuous-time T–S fuzzy systems. With the rapid development of the digital signal processor, the representation of discrete-time systems and discrete-time control approaches play more important roles in the control theory. However, there are few results on SMC for discrete-time T–S fuzzy systems. Motivated by these, we desire to extend the result to the discrete-time fuzzy MJSs with uncertainties. Now, considering the following fuzzy MJSs:
Plant rule $i^{\alpha(k)}$: IF $\rho_1(k)$ is $\varphi_{i1}^{\alpha(k)}$, and ..., and $\rho_q(k)$ is $\varphi_{iq}^{\alpha(k)}$, THEN

$$\begin{cases} x(k+1) = A_{i\alpha(k)}x(t) + B_{i\alpha(k)}(u(k) + f(x(k))) + C_{i\alpha(k)}w(k), \\ \quad z(k) = D_{i\alpha(k)}x(k) + E_{i\alpha(k)}w(k), \end{cases} \tag{7.34}$$

where $x(k) \in R^{n_x}$, $u(k) \in R^{n_u}$, $w(k) \in R^{n_w}$ and $z(k) \in R^{n_z}$ are the state, the control input, the noise belonging to $l_2[0, +\infty)$ and the output, respectively. i ($i \in \mathcal{R} = \{1, \ldots, r\}$) is the ith fuzzy rule. $\rho_1(k), \ldots, \rho_q(k)$ are the premise variables and $\varphi_{i1}^{\alpha(k)}, \ldots, \varphi_{iq}^{\alpha(k)}$ are the fuzzy sets. The nonlinear term $f(x(k))$ denotes the matched uncertainty with $\|f(x(k))\| < \kappa\|x(k)\|$, where κ is known. The stochastic variable $\alpha(k)$ ($\alpha(k) \in \mathcal{M}_1 = \{1, \ldots, M_1\}$) is used to represent the discrete-time Markov process, subject to the transition probability matrix $\Theta = [\theta_{ab}]$ and

$$\Pr\{\alpha(k+1) = b|\alpha(k) = a\} = \theta_{ab}, \tag{7.35}$$

where $0 \leq \theta_{ab} \leq 1$ and $\sum_{b=1}^{M_1} \theta_{ab} = 1$.

The overall fuzzy systems from (7.34) with $\alpha(k) = a$ can be obtained as

$$\begin{cases} x(k+1) = A_{ha}(k)x(k) + B_{ha}(k)(u(k) + f(x(k))) \\ \qquad\qquad + C_{ha}(k)w(k), \\ \quad z(k) = D_{ha}(k)x(k) + E_{ha}(k)w(k), \end{cases} \tag{7.36}$$

where $B_{ha}(k)$ is assumed to be full column rank, i.e., $rank(B_{ha}(k)) = n_u$. And

$$A_{ha}(k) = \sum_{i=1}^{r} h_{ia}(k)A_{ia}, \ B_{ha}(k) = \sum_{i=1}^{r} h_{ia}(k)B_{ia},$$

$$C_{ha}(k) = \sum_{i=1}^{r} h_{ia}(k)C_{ia}, \ D_{ha}(k) = \sum_{i=1}^{r} h_{ia}(k)D_{ia},$$

$$E_{ha}(k) = \sum_{i=1}^{r} h_{ia}(k)E_{ia}, \ \sum_{i=1}^{r} h_{ia}(k) = 1,$$

$$h_{ia}(k) = \frac{\prod_{j=1}^{q} \varphi_{ij}^{a}(\rho_j(k))}{\sum_{i=1}^{r} \prod_{j=1}^{q} \varphi_{ij}^{a}(\rho_j(k))} \geq 0.$$

$h_{ia}(k)$ denotes the normalized membership function and $\varphi_{ij}^{a}(\rho_j(k))$ is the the grade of membership of $\rho_j(k)$ in φ_{ij}^{a}.

The same aim as that in continuous-time fuzzy systems is considered, that is, developing an appropriate SMC strategy to ensure the stochastic stability of the sliding mode dynamics with dissipative performance.

We firstly use $\beta(k)$ ($\beta(k) \in \mathcal{M}_2 = \{1, \ldots, M_2\}$) to detect mode $\alpha(k)$ and the corresponding HMM is represented as

$$\Pr\{\beta(k) = c | \alpha(k) = a\} = \varrho_{ac}, \tag{7.37}$$

where the conditional transition probability matrix is $\Gamma = [\varrho_{ac}]$ with $\varrho_{ac} \geq 0$ and $\sum_{c=1}^{M_2} \varrho_{ac} = 1$. Then, we construct the discrete-time AFISMC surface with $\beta(k) = c$ as

$$\begin{aligned} s(k) = B_{ha}^{T}(0)x(0) &+ \sum_{\upsilon=1}^{k} B_{ha}^{T}(\upsilon - 1)x(\upsilon) \\ &- \sum_{\upsilon=1}^{k-1} B_{ha}^{T}(\upsilon)(A_{ha}(\upsilon) + B_{ha}(\upsilon)K_{hc}(\upsilon))x(\upsilon). \end{aligned} \tag{7.38}$$

Here, $K_{hc}(k) = \sum_{j=1}^{r} h_{jc}(k)K_{jc}$ and K_{jc} is the controller gain to be solved. It further derives that

$$\begin{aligned} s(k+1) = s(k) + B_{ha}^{T}(k)x(k+1) &- B_{ha}^{T}(k)(A_{ha}(k) \\ &+ B_{ha}(k)K_{hc}(k))x(k). \end{aligned} \tag{7.39}$$

The ideal specified sliding surface is: for $k > k^* > 0$, $s(k+1) = s(k) = 0$. In this case, we have the equivalent controller, as follows:

$$u_{eq}(k) = K_{hc}(k)x(k) - f(x(k)) - (B_{ha}^{T}(k)B_{ha}(k))^{-1}B_{ha}^{T}(k)C_{ha}(k)w(k). \tag{7.40}$$

By applying the control input $u_{eq}(k)$, the sliding mode dynamics are derived as

$$\begin{cases} x(k+1) = \hat{A}_{hac}(k)x(k) + \tilde{B}_{ha}(k)C_{ha}(k)w(k), \\ \quad z(k) = D_{ha}(k)x(k) + E_{ha}(k)w(k), \end{cases} \tag{7.41}$$

where

$$\hat{A}_{hac}(k) = A_{ha}(k) + B_{ha}(k)K_{hc}(k),$$
$$\tilde{B}_{ha}(k) = I - B_{ha}(k)(B_{ha}^T(k)B_{ha}(k))^{-1}B_{ha}^T(k).$$

Remark 7.2 By using the sliding surface (7.39), we can clearly observe that the matched uncertainty in (7.36) is completely compensated in the sliding mode dynamics (7.41) and the effect of external noise $w(k)$ is not amplified as well, which show the strong robustness of the proposed discrete-time AFISMC approach. In addition, the developed continuous-time AFISMC scheme also shows the same advantages, which can be found in (7.8).

In the following, we analyze two problems: (i) the solution to the controller gain K_{jc}, which can also guarantee the stochastic stability of system (7.41) with a desired dissipative performance; and (ii) the synthesis on the discrete-time AFISMC law to ensure that trajectories of system (7.41) operate on a bounded region around the surface $s(k) = 0$ and remain there later.

Theorem 7.6 *System (7.41) is stochastically stable with a desired dissipative performance* (G_1, G_2, G_3, δ) *if there exist matrices* $P_a > 0$, $W_{ac} > 0$, K_{jc}, *scalar* $\varepsilon > 0$ *for all* $i, j \in \mathcal{R}$, $a, b \in \mathcal{M}_1$ *and* $c \in \mathcal{M}_2$ *satisfying*

$$\sum_{c=1}^{M_2} \varrho_{ac} W_{ac} < P_a, \tag{7.42}$$

$$\Psi_{iiac} < 0, \tag{7.43}$$

$$\Psi_{ijac} + \Psi_{jiac} < 0, \; i < j, \tag{7.44}$$

where

$$\Psi_{ijac} = \begin{bmatrix} -W_{ac} & * & * & * & * & * \\ -G_2^T D_{ia} & \Psi_{ia}^{22} & * & * & * & * \\ G_1^+ D_{ia} & G_1^+ E_{ia} & -I & * & * & * \\ \Psi_{ijac}^{41} & 0 & 0 & \Psi^{44} & * & * \\ 0 & 0 & 0 & \Psi^{54} & -\varepsilon I & * \\ 0 & \varepsilon C_{ia} & 0 & 0 & 0 & -\varepsilon I \end{bmatrix},$$

$$\Psi_{ia}^{22} = -E_{ia}^T G_2 - G_2^T E_{ia} - G_3 + \delta I,$$
$$\Psi_{ijac}^{41} = [\sqrt{\theta_{a1}}(A_{ia} + B_{ia}K_{jc})^T P_1,$$
$$\ldots, \sqrt{\theta_{aM_1}}(A_{ia} + B_{ia}K_{jc})^T P_{M_1}]^T,$$

$$\Psi^{44} = -diag\{P_1, \ldots, P_{M_1}\},$$
$$\Psi^{54} = -[\sqrt{\theta_{a1}} P_1, \ldots, \sqrt{\theta_{aM_1}} P_{M_1}].$$

Proof Lyapunov function is selected as

$$V(k) = x^T(k) P_{\alpha(k)} x(k) \tag{7.45}$$

with $P_{\alpha(k)} > 0$. Following the similar proof steps in [8] and Theorem 7.1, we can obtain Theorem 7.6. Accordingly, the proof is omitted.

The nonlinear coupling between P_a and K_{jc} also exists in discrete-time fuzzy systems, and we develop a more convenient algorithm, as follows:

Theorem 7.7 *System (7.41) is stochastically stable with a desired dissipative performance* (G_1, G_2, G_3, δ) *if there exist matrices* $\bar{P}_a > 0$, $\bar{W}_{ac} > 0$, X, Y_{jc}, *scalar* $\varepsilon > 0$ *for all* $i, \ j \in \mathcal{R}, a, \ b \in \mathcal{M}_1$ *and* $c \in \mathcal{M}_2$ *satisfying*

$$\begin{bmatrix} -\bar{P}_a & * \\ \mathcal{P}_a & -\mathcal{W}_a \end{bmatrix} < 0, \tag{7.46}$$

$$\Lambda_{iiac} < 0, \tag{7.47}$$

$$\Lambda_{ijac} + \Lambda_{jiac} < 0, \ i < j, \tag{7.48}$$

where

$$\mathcal{P}_a = [\varrho_{a1}^{0.5} I, \ldots, \varrho_{aM_2}^{0.5} I]^T \bar{P}_a, \ \bar{P}_a = P_a^{-1},$$
$$\mathcal{W}_a = diag\{\bar{W}_{a1}, \ldots, \bar{W}_{aM_2}\}, \ \bar{W}_{ac} = W_{ac}^{-1},$$
$$\Lambda_{ijac} = \begin{bmatrix} \Lambda^{11} & * & * & * & * & * \\ -G_2^T D_{ia} X & \Lambda_{ia}^{22} & * & * & * & * \\ G_1^+ D_{ia} X & G_1^+ E_{ia} & -I & * & * & * \\ \Lambda_{ijac}^{41} & 0 & 0 & \Lambda^{44} & * & * \\ 0 & 0 & 0 & \Lambda^{54} & -\varepsilon I & * \\ 0 & \varepsilon C_{ia} & 0 & 0 & 0 & -\varepsilon I \end{bmatrix},$$
$$\Lambda^{11} = \bar{W}_{ac} - X - X^T,$$
$$\Lambda_{ia}^{22} = -E_{ia}^T G_2 - G_2^T E_{ia} - G_3 + \delta I,$$
$$\Lambda_{ijac}^{41} = [\sqrt{\theta_{a1}} (A_{ia} X + B_{ia} Y_{jc})^T, \ldots, \sqrt{\theta_{aM_1}} (A_{ia} X + B_{ia} Y_{jc})^T]^T,$$
$$\Lambda^{44} = -diag\{\bar{P}_1, \ldots, \bar{P}_{M_1}\},$$
$$\Lambda^{54} = [\sqrt{\theta_{a1}} I, \ldots, \sqrt{\theta_{aM_1}} I].$$

Moreover, the controller gain can be computed from

$$K_{jc} = Y_{jc} X^{-1}. \tag{7.49}$$

Proof Theorem 7.7 is achieved by using the similar proof process in [8].

Remark 7.3 The used Lyapunov function matrix $P_{\alpha(t)}$ ($P_{\alpha(k)}$) depends on mode $\alpha(t)$ ($\alpha(k)$) of original systems, which can flexibly adapt to mode $\alpha(t)$ ($\alpha(k)$) and bring less conservatism than the common Lyapunov function matrix. On the other hand, since $P_{\alpha(t)}$ ($P_{\alpha(k)}$) only relies on $\alpha(t)$ ($\alpha(k)$) and has no relation with $\beta(t)$ ($\beta(k)$), it may lead to some conservatism. To further reduce conservatism, it is meaningful to investigate how to design the Lyapunov function matrix depending on both $\alpha(t)$ ($\alpha(k)$) and $\beta(t)$ ($\beta(k)$) for the asynchronous controller design problem in future. It is worth pointing out that inequalities in Theorem 7.2 is nonlinear and nonlinearity exists between γ and $\bar{P}_a$. We firstly need to find a feasible γ for inequalities and then adopt LMI Toolbox in Matlab to compute the controller gains. In addition, we can obtain the optimal dissipative performance δ^* by minimizing $-\delta$ in inequalities in Theorems 7.1, 7.2 and 7.6, 7.7, respectively.

We design the following AFISMC law to ensure that system trajectories (7.36) can operate in the sliding mode region.

Theorem 7.8 *The system trajectories (7.36) can be forced into a small bounded region via the following AFISMC law:*

$$u(k) = K_{hc}(k)x(k) - \mu s(k) + \chi(k) \cdot o(s(k)), \tag{7.50}$$

where

$$o(s(k)) = \begin{cases} -\dfrac{\hat{B}_{ha}(k)s(k)}{\|\hat{B}_{ha}(k)s(k)\|}, & \|\hat{B}_{ha}(k)s(k)\| > o, \\ -\dfrac{\hat{B}_{ha}(k)s(k)}{o}, & \|\hat{B}_{ha}(k)s(k)\| \leq o, \end{cases}$$

$$\chi(k) = (\kappa\chi_1 + \chi_2)/\chi_3,\ \chi_1 = \|\hat{B}_{ha}(k)\| \|x(k)\|,\ \mu > 0,$$

$$\chi_2 = \|B_{ha}^T(k) C_{ha}(k) w(k)\|,\ \chi_3 = \sqrt{\lambda_{min}(\hat{B}_{ha}(k)\hat{B}_{ha}(k))},$$

$$\hat{B}_{ha}(k) = B_{ha}^T(k) B_{ha}(k).$$

Proof The candidate Lyapunov function is

$$V(k) = 0.5 s^T(k)s(k). \tag{7.51}$$

Then we obtain

$$\begin{aligned} \Delta V(k) &= V(k+1) - V(k) \\ &= s^T(k)\Delta s(k) + \tau_0, \end{aligned} \tag{7.52}$$

where $\tau_0 = \frac{1}{2}\Delta s^T(k)\Delta s(k)$ with $\Delta s(k) = s(k+1) - s(k)$. From (7.39) and (7.50), it follows that

$$\begin{aligned}\Delta s(k) = & -\mu\hat{B}_{ha}(k)s(k) + \hat{B}_{ha}(k)f(x(k)) \\ & + B_{ha}^T(k)C_{ha}(k)w(k) + \chi(k)\hat{B}_{ha}(k)\cdot o(s(k)).\end{aligned} \tag{7.53}$$

If $\|\hat{B}_{ha}(k)s(k)\| > o$,

$$\begin{aligned}E\{\Delta V(k)\} < & -\mu s^T(k)\hat{B}_{ha}(k)s(k) - \chi(k)\|\hat{B}_{ha}(k)s(k)\| \\ & + \tau_0 + \|s(k)\|(\kappa\chi_1 + \chi_2\|) \\ < & -\mu s^T(k)\hat{B}_{ha}(k)s(k) - \chi(k)\chi_3\|s(k)\| \\ & + \tau_0 + \|s(k)\|(\kappa\chi_1 + \chi_2\|) \\ = & -\mu s^T(k)\hat{B}_{ha}(k)s(k) + \tau_0.\end{aligned} \tag{7.54}$$

If $\|\hat{B}_{ha}(k)s(k)\| < o$,

$$\begin{aligned}E\{\Delta V(k)\} < & -\mu s^T(k)\hat{B}_{ha}(k)s(k) - \frac{\chi(k)}{o}\|\hat{B}_{ha}(k)s(k)\|^2 \\ & + \tau_0 + \|s(k)\|(\kappa\chi_1 + \chi_2\|) \\ < & -\mu s^T(k)\hat{B}_{ha}(k)s(k) - \frac{\chi(k)}{o}\chi_3^2\|s(k)\|^2 \\ & + \tau_0 + \|s(k)\|(\kappa\chi_1 + \chi_2) \\ = & -\mu s^T(k)\hat{B}_{ha}(k)s(k) \\ & + \chi(k)\chi_3(1 - \frac{\chi_3}{o}\|s(k)\|)\|s(k)\| + \tau_0.\end{aligned} \tag{7.55}$$

We can select an appropriate μ to ensure that $E\{\Delta V(k)\} < 0$ and $\|s(k)\| \leq o$. Then, state trajectories (7.36) operate in a bounded region around $s(k) = 0$ in limited time.

In the following, a discrete-time adaptive AFISMC law is proposed to deal with a situation that κ is unknown.

Theorem 7.9 *The system trajectories (7.36) can be forced into a small bounded region via the following AFISMC law:*

$$u(k) = K_{hc}(k)x(k) - \mu s(k) + \hat{\chi}(k)\cdot o(s(k)), \tag{7.56}$$

where

$$o(s(k)) = \begin{cases} -\dfrac{\hat{B}_{ha}(k)s(k)}{\|\hat{B}_{ha}(k)s(k)\|}, & \|\hat{B}_{ha}(k)s(k)\| > o, \\ -\dfrac{\hat{B}_{ha}(k)s(k)}{o}, & \|\hat{B}_{ha}(k)s(k)\| \leq o, \end{cases}$$

$$\hat{\chi}(k) = (\hat{\kappa}(k)\chi_1 + \chi_2)/\chi_3,\ \chi_1 = \|\hat{B}_{ha}(k)\|\|x(k)\|,\ \mu > 0,$$

$$\chi_2 = \|B_{ha}^T(k)C_{ha}(k)w(k)\|,\ \chi_3 = \sqrt{\lambda_{min}(\hat{B}_{ha}(k)\hat{B}_{ha}(k))},$$

$$\hat{B}_{ha}(k) = B_{ha}^T(k)B_{ha}(k).$$

The estimation $\hat{\kappa}(k)$ satisfies $\Delta\hat{\kappa}(k) = \hat{\kappa}(k+1) - \hat{\kappa}(k) = \varpi\|\hat{B}_{ha}(k)\|\|s(k)\|\|x(k)\|$ with $\hat{\kappa}(0) = 0$ and $\varpi > 0$.

Proof The candidate Lyapunov function is

$$V(k) = 0.5s^T(k)s(k) + \frac{1}{2\varpi}\tilde{\kappa}^2(k), \tag{7.57}$$

where $\tilde{\kappa}(k) = \hat{\kappa}(k) - \kappa$. Then we obtain

$$\begin{aligned} \Delta V(k) &= V(k+1) - V(k) \\ &= s^T(k)\Delta s(k) + \frac{1}{\varpi}\tilde{\kappa}(k)\Delta\tilde{\kappa}(k) + \tau_0 \end{aligned} \tag{7.58}$$

where $\tau_0 = \frac{1}{2}\Delta s^T(k)\Delta s(k) + \frac{1}{2\varpi}\Delta\tilde{\kappa}^2(k)$, $\Delta s(k) = s(k+1) - s(k)$ and $\Delta\tilde{\kappa}(k) = \tilde{\kappa}(k+1) - \tilde{\kappa}(k)$. By similar proof lines in Theorem 7.8, we obtain Theorem 7.9. Hence, the proof is omitted.

7.5 Illustrative Example

In this section, another single-link rigid robot system [4, 9] is used to demonstrate the feasibility and the effectiveness of developed approaches. Its mathematical model is given as

$$J_{\alpha(t)}\ddot{\vartheta} = -(0.5m_{1\alpha(t)} + m_{2\alpha(t)})glsin(\vartheta) + u(t) + w(t),$$

where ϑ ($\vartheta \in [0, 0.5\pi]$) is the joint rotation angle. l denotes the length of the robot link with $l = 0.5$. g represents the gravity acceleration with $g = 9.8$. $u(t)$ is the control input. In this example, we assume that the system is subject to the external noise $w(t)$ and parameters' changes, namely, the masses of the load $m_{1\alpha(t)}$ and the rigid link $m_{2\alpha(t)}$ are varying, which are modelled by the stochastic Markov process $\alpha(t)$. $J_{\alpha(t)}$ is the moment of inertia with $J_{\alpha(t)} = \frac{1}{3}m_{1\alpha(t)}l^2 + m_{2\alpha(t)}l^2$.

By using the T–S model approach with $x_1(t) = \vartheta$ and $x_2(t) = \dot{\vartheta}$, the original systems are modelled as

Plant rule $1^{\alpha(t)}$: IF $x_1(t)$ is about 0 rad, THEN

$$\begin{cases} \dot{x}(t) = A_{1\alpha(t)}x(t) + B_{1\alpha(t)}(u(t) + f(x(t))) + C_{1\alpha(t)}w(t), \\ z(t) = D_{1\alpha(t)}x(t) + E_{1\alpha(t)}w(t), \end{cases}$$

Plant rule $2^{\alpha(t)}$: IF $x_1(t)$ is about 0.5π rad, THEN

$$\begin{cases} \dot{x}(t) = A_{2\alpha(t)}x(t) + B_{2\alpha(t)}(u(t) + f(x(t))) + C_{2\alpha(t)}w(t), \\ z(t) = D_{2\alpha(t)}x(t) + E_{2\alpha(t)}w(t), \end{cases}$$

where

$$\begin{aligned} A_{1\alpha(t)} &= \begin{bmatrix} 0 & 1 \\ -\frac{(0.5m_{1\alpha(t)}+m_{2\alpha(t)})gl}{J_{\alpha(t)}} & 0 \end{bmatrix}, \; C_{1\alpha(t)} = \begin{bmatrix} 0 \\ 1 \end{bmatrix}, \\ A_{2\alpha(t)} &= \begin{bmatrix} 0 & 1 \\ -\frac{(m_{1\alpha(t)}+2m_{2\alpha(t)})gl}{\pi J_{\alpha(t)}} & 0 \end{bmatrix}, \; C_{2\alpha(t)} = \begin{bmatrix} 0 \\ 1 \end{bmatrix}, \\ B_{11} &= B_{21} = \begin{bmatrix} 0 \\ \frac{1}{J_1} \end{bmatrix}, \; B_{12} = B_{22} = \begin{bmatrix} 0 \\ \frac{1}{J_2} \end{bmatrix}, \\ D_{1\alpha(t)} &= D_{2\alpha(t)} = \begin{bmatrix} 1 & 0 \end{bmatrix}, \; E_{1\alpha(t)} = E_{2\alpha(t)} = 1, \\ m_{11} &= 1.5, \; m_{12} = 1.2, \; m_{21} = 3, \; m_{22} = 3.1, \\ f(x(t)) &= 0.11e^{-t}sin(t)x_1(t), \; \alpha(t) = \{1, 2\}. \end{aligned}$$

The normalized membership functions are plotted in Fig. 7.1. It is supposed that $\alpha(t)$ and $\beta(t)$ obey the following transition probability matrices:

$$\Theta = \begin{bmatrix} -2 & 2 \\ 3 & -3 \end{bmatrix}, \; \Gamma = \begin{bmatrix} 0.2 & 0.4 & 0.4 \\ 0.4 & 0.3 & 0.3 \end{bmatrix}.$$

The dissipative performance is given as $G_1 = -1$, $G_2 = 1$ and $G_3 = 2$. By computing, we find that the controller gains (7.59) can make the inequalities in Theorem 7.1 feasible and the dissipative performance $\delta = 2.4486$ is obtained.

$$\begin{aligned} K_{11} &= [10.4991 \; -5.1057], \; K_{21} = [3.8800 \; -5.0628], \\ K_{12} &= [10.8456 \; -5.6570], \; K_{22} = [3.900 \; -5.1648], \\ K_{13} &= [11.5143 - 5.2057], \; K_{23} = [3.8912 \; -5.5628]. \end{aligned} \tag{7.59}$$

We assume the initial condition is $x(0) = [\pi/3 \; 5\pi/12]^T$ and the noise is $w(t) = 5e^{-0.08t}$. With $\mu = 8$, $\kappa = 0.11$ and $o = 0.001$, the state responses of the sliding mode dynamics are shown in Fig. 7.2 and the sliding surface is plotted in Fig. 7.3.

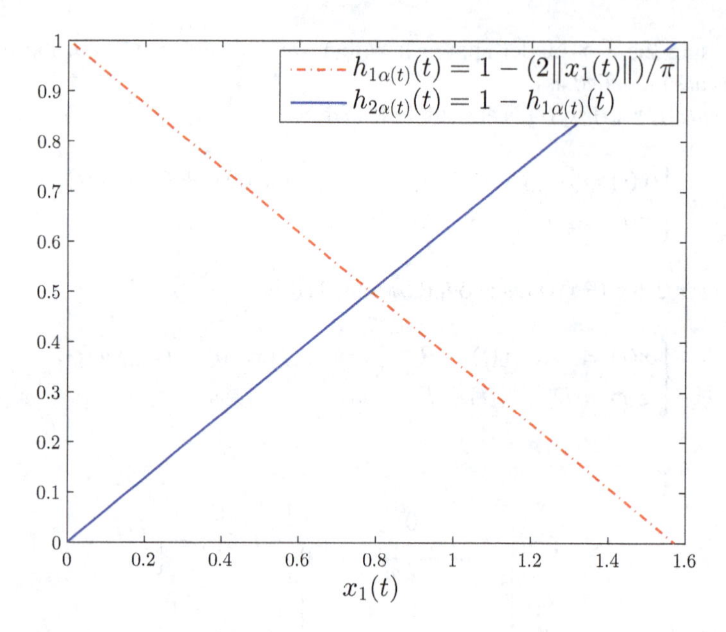

Fig. 7.1 Normalized membership functions

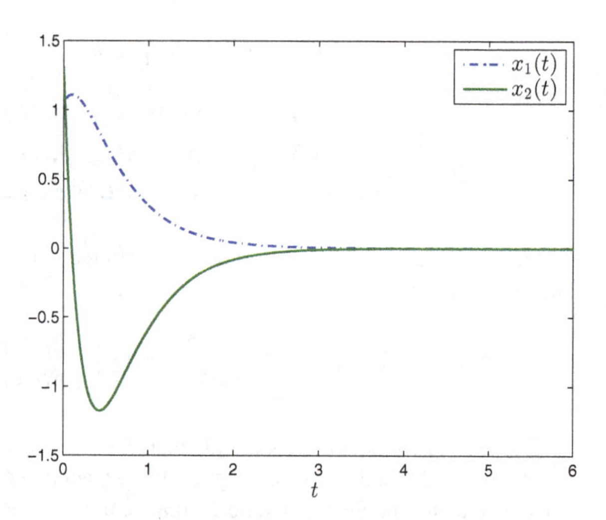

Fig. 7.2 State trajectories of sliding mode dynamics (7.8)

Then, with the sampling period $T = 0.025$, we adopt the first-order Euler approximation approach to discretize the continuous-time fuzzy MJSs, and this discretization approach has been employed in fault detection problem [10] and H_∞ control problem [11]. The discrete-time fuzzy MJSs are obtained as *Plant rule* $1^{\alpha(k)}$: IF $x_1(k)$ is about 0 rad, THEN

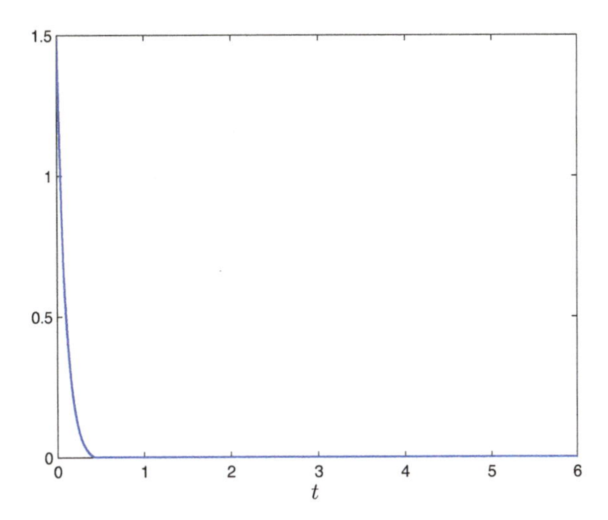

Fig. 7.3 Sliding surface (7.5)

$$\begin{cases} x(k+1) = A_{1\alpha(k)}x(k) + B_{1\alpha(k)}(u(k) + f(x(k))) \\ \qquad + C_{1\alpha(k)}w(k), \\ z(k) = D_{1\alpha(k)}x(k) + E_{1\alpha(k)}w(k), \end{cases}$$

Plant rule $2^{\alpha(k)}$: IF $x_1(k)$ is about 0.5π rad, THEN

$$\begin{cases} x(k+1) = A_{2\alpha(k)}x(k) + B_{2\alpha(k)}(u(k) + f(x(k))) \\ \qquad + C_{2\alpha(k)}w(k), \\ z(k) = D_{2\alpha(k)}x(k) + E_{2\alpha(k)}w(k), \end{cases}$$

where

$$\begin{aligned}
&A_{1\alpha(k)} = \begin{bmatrix} 1 & T \\ -\frac{(0.5m_{1\alpha(k)}+m_{2\alpha(k)})glT}{J_{\alpha(k)}} & 1 \end{bmatrix}, \; C_{1\alpha(k)} = \begin{bmatrix} 0 \\ T \end{bmatrix}, \\
&A_{2\alpha(k)} = \begin{bmatrix} 1 & T \\ -\frac{(m_{1\alpha(k)}+2m_{2\alpha(k)})glT}{\pi J_{\alpha(k)}} & 1 \end{bmatrix}, \; C_{1\alpha(k)} = \begin{bmatrix} 0 \\ T \end{bmatrix}, \\
&B_{11} = B_{21} = \begin{bmatrix} 0 \\ \frac{T}{J_1} \end{bmatrix}, \; B_{12} = B_{22} = \begin{bmatrix} 0 \\ \frac{T}{J_2} \end{bmatrix}, \\
&D_{1\alpha(t)} = D_{2\alpha(k)} = \begin{bmatrix} 1 \; 0 \end{bmatrix}, \; E_{1\alpha(k)} = E_{2\alpha(k)} = 1, \\
&m_{11} = 1.5, \; m_{12} = 1.2, \; m_{21} = 3, \; m_{22} = 3.1, \\
&h_{1\alpha(k)}(k) = 1 - (2\|x_1(k)\|)/\pi, \; h_{2\alpha(k)}(k) = 1 - h_{1\alpha(k)}(k), \\
&f(x(k)) = 0.11e^{-k}sin(k)x_1(k), \; \alpha(k) = \{1, 2\}.
\end{aligned}$$

The probability matrices of Θ and Γ are assumed to be:

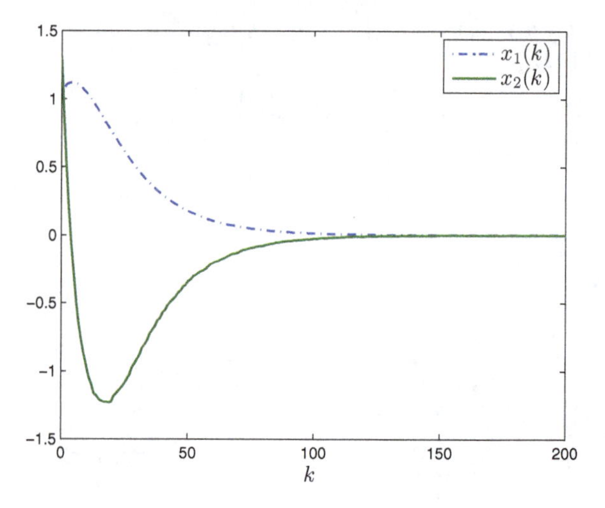

Fig. 7.4 State trajectories of sliding mode dynamics (7.41)

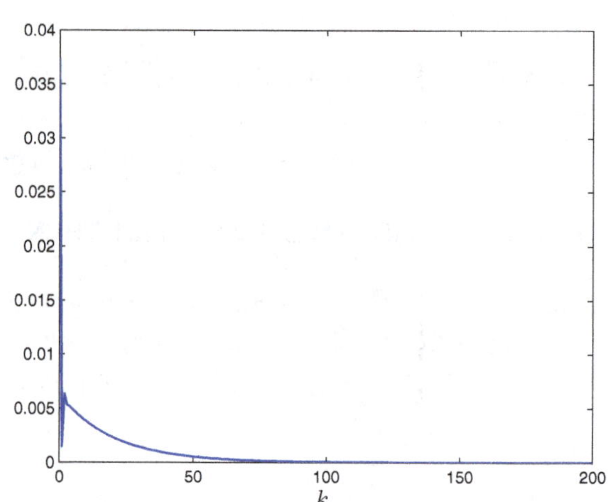

Fig. 7.5 Sliding surface (7.39)

$$\Theta = \begin{bmatrix} 0.8 & 0.2 \\ 0.3 & 0.7 \end{bmatrix}, \ \Gamma = \begin{bmatrix} 0.2 & 0.4 & 0.4 \\ 0.4 & 0.3 & 0.3 \end{bmatrix}.$$

Let $G_1 = -1$, $G_2 = 1$ and $G_3 = 2$. There is a feasible solution in Theorem 7.6 when controller gains are taken as (7.60) and $\delta = 1.8463$ with

$$\begin{aligned} K_{11} &= [11.2312 \ \ -5.0207], \ K_{21} = [3.2605 \ \ -4.9628], \\ K_{12} &= [9.7901 \ \ -6.6571], \ K_{22} = [4.9150 \ \ -5.0628], \\ K_{13} &= [10.5346 \ \ -5.0074], \ K_{23} = [3.9872 \ \ -5.6678]. \end{aligned} \tag{7.60}$$

Given the same initial state and the noise $w(k) = 10e^{-0.05k}$, Fig. 7.4 shows the state response of the sliding mode dynamics and Fig. 7.5 plots the sliding surface with $o = 0.005$ and $\mu = 1400$. From these results, we can clearly conclude that our proposed AFISMC approaches for fuzzy MJSs are feasible and effective.

7.6 Conclusion

In this chapter, the problems of asynchronous SMC for both continuous-time and discrete-time fuzzy MJSs have been investigated. New continuous-time and discrete-time AFISMC surfaces have been constructed with the observed jump mode via HMMs, respectively. Sufficient conditions for the performance of sliding mode dynamics have been proposed, which give the solution to the sliding mode controller gains as well. Then, the corresponding AFISMC law and an adaptive one for the unknown bound of uncertainties have been constructed to force the system onto a small bounded region and make it work there later. We have verified the solvability and validity of the proposed approaches via an example simulation.

References

1. Song, J., Niu, Y., Zou, Y.: Finite-time stabilization via sliding mode control. IEEE Trans. Autom. Control **62**(3), 1478–1483 (2017)
2. Wu, L., Gao, Y., Liu, J., Li, H.: Event-triggered sliding mode control of stochastic systems via output feedback. Automatica **82**, 79–92 (2017)
3. Li, J., Zhang, Q., Yan, X.-G., Spurgeon, S.K.: Robust stabilization of T-S fuzzy stochastic descriptor systems via integral sliding modes. IEEE Trans. Cybern. **48**(9), 2736–2749 (2018)
4. Xue, Y., Zheng, B.-C., Yu, X.: Robust sliding mode control for T-S fuzzy systems via quantized state feedback. IEEE Trans. Fuzzy Syst. **26**(4), 2261–2272 (2018)
5. Wang, Y., Shen, H., Karimi, H.R., Duan, D.: Dissipativity-based fuzzy integral sliding mode control of continuous-time T-S fuzzy systems. IEEE Trans. Fuzzy Syst. **26**(3), 1164–1176 (2018)
6. Gao, H., Chen, T.: H_∞ estimation for uncertain systems with limited communication capacity. IEEE Trans. Autom. Control **52**(11), 2070–2084 (2007)
7. Yoo, D.S., Chung, M.J.: A variable structure control with simple adaptation laws for upper-bounds on the norm of the uncertainties. IEEE Trans. Autom. Control **37**(6), 860–864 (1992)
8. Wu, Z.-G., Dong, S., Su, H., Li, C.: Asynchronous dissipative control for fuzzy Markov jump systems. IEEE Trans. Cybern. **48**(8), 2426–2436 (2018)
9. Zhang, H., Yang, J., Su, C.-Y.: T-S fuzzy-model-based robust H_∞ design for networked control systems with uncertainties. IEEE Trans. Ind. Inform. **3**(4), 289–301 (2007)
10. Li, F., Shi, P., Lim, C.-C., Wu, L.: Fault detection filtering for nonhomogeneous Markovian jump systems via a fuzzy approach. IEEE Trans. Fuzzy Syst. **26**(1), 131–141 (2018)
11. Zhang, L., Ning, Z., Shi, P.: Input-output approach to control for fuzzy Markov jump systems with time-varying delays and uncertain packet dropout rate. IEEE Trans. Cybern. **45**(11), 2449–2460 (2015)

Chapter 8
Filtering for Discrete-Time Switched Fuzzy Systems with Quantization

8.1 Introduction

In this chapter, the H_∞ and L_2–L_∞ filtering problems are investigated for discrete-time switched fuzzy systems with quantization. We apply the sector bound approach to deal with quantization errors as sector bound uncertainties. By utilizing the fuzzy-basis-dependent Lyapunov function, sufficient conditions are given such that the filtering error system is stochastically stable with the prescribed H_∞ or L_2–L_∞ performance index. Slack matrices are introduced to eliminate the coupling between the Lyapunov matrices and the system matrices. And main results are expressed as LMIs.

8.2 Preliminary Analysis

The problems of H_∞ and L_2–L_∞ filtering are shown in Fig. 8.1. The studied switched nonlinear systems are represented by the T–S fuzzy model, which have two level functions, namely, the switching function and the fuzzy function. Due to digital communication limits, the signal y_{fk} received by the filter is no equivalent to the output y_k and we take the quantization effect into consideration by the classical sector bound approach. In the following, we model the whole problem by mathematical methods.

8.2.1 Switched Fuzzy Systems

Consider the following discrete-time switched T–S fuzzy system:
Plant rule i: IF θ_{1k} is M_{i1}, θ_{2k} is M_{i2}, ..., and θ_{pk} is M_{ip}, THEN

$$\begin{cases} x_{k+1} = A_{\tau_k i} x_k + B_{\tau_k i} w_k, \\ y_k = C_{\tau_k i} x_k + D_{\tau_k i} w_k, \\ z_k = E_{\tau_k i} x_k, \end{cases} \tag{8.1}$$

S. Dong et al., *Control and Filtering of Fuzzy Systems with Switched Parameters*, Studies in Systems, Decision and Control 268,
https://doi.org/10.1007/978-3-030-35566-1_8

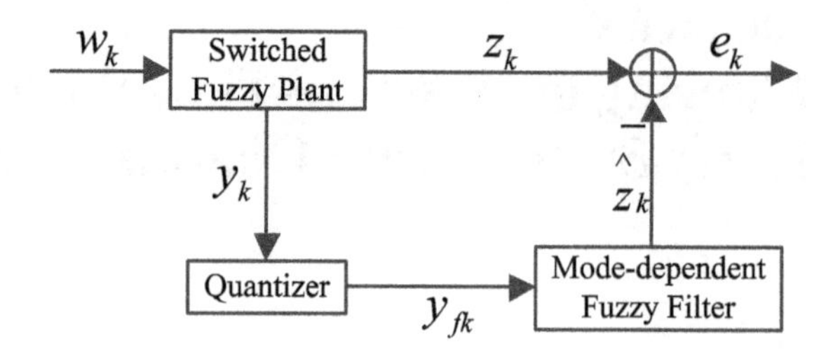

Fig. 8.1 Filtering design with quantization

where M_{ij} is a fuzzy set, and θ_{jk} is the premise variable ($i \in \{1, 2, \ldots, r\}$, $j \in \{1, 2, \ldots, p\}$); r is the number of IF-THEN rules; r, p are positive integers; $x_k \in R^n$ is the state; $w_k \in R^m$ is the disturbance that belongs to $l_2[0, \infty)$; $y_k \in R^q$ is the measurement output; $z_k \in R^v$ is the signal to be estimated. $A_{\tau_k i}$, $B_{\tau_k i}$, $C_{\tau_k i}$, $D_{\tau_k i}$ and $E_{\tau_k i}$ are known matrices with appropriate dimensions. The stochastic variable $\tau_k \in \{1, 2, 3, \ldots, L\}$ is used to describe the switching phenomenon satisfying

$$\Pr\{\tau_k = l\} = \pi_l, \ \pi_l \geq 0, \ \text{and} \sum_{l=1}^{L} \pi_l = 1. \tag{8.2}$$

A set of stochastic variables π_{kl} are denoted by $\pi_{kl} = \begin{cases} 1, & \tau_k = l, \\ 0, & \tau_k \neq l. \end{cases}$ It follows that $E\{\pi_{kl}\} = \pi_l$.

By using a center-average defuzzifier, the product inference and the singleton fuzzifier, system (8.1) is described as

$$\begin{cases} x_{k+1} = A_{lh} x_k + B_{lh} w_k, \\ y_k = C_{lh} x_k + D_{lh} w_k, \\ z_k = E_{lh} x_k, \end{cases} \tag{8.3}$$

where

$$\begin{aligned} A_{lh} &= \sum_{i=1}^{r} h_i(\theta_k) A_{li}, \ B_{lh} = \sum_{i=1}^{r} h_i(\theta_k) B_{li}, \\ C_{lh} &= \sum_{i=1}^{r} h_i(\theta_k) C_{li}, \ D_{lh} = \sum_{i=1}^{r} h_i(\theta_k) D_{li}, \\ E_{lh} &= \sum_{i=1}^{r} h_i(\theta_k) E_{li}, \ h \triangleq (h_1(\theta_k), h_2(\theta_k), \cdots, h_r(\theta_k)), \\ \theta_k &= [\theta_{1k}, \theta_{2k}, \ldots, \theta_{pk}]. \end{aligned}$$

The normalized fuzzy weighting functions are defined as

$$h_i(\theta_k) = \frac{\prod_{j=1}^{p} M_{ij}(\theta_{jk})}{\sum_{i=1}^{r} \prod_{j=1}^{p} M_{ij}(\theta_{jk})}, \tag{8.4}$$

in which $M_{ij}(\theta_{jk})$ is the grade of the membership of θ_{jk} in M_{ij}. We describe $h_i(\theta_k)$ as h_i for brevity in the following. It is necessary to mention that the normalized fuzzy weighting functions satisfy

$$\begin{cases} h_i \geq 0,\ i = \{1, 2, ...r\}, \\ \sum_{i=1}^{r} h_i = 1. \end{cases} \tag{8.5}$$

8.2.2 *Measurement Quantization*

In the chapter, the measured output is assumed to be quantized by the logarithmic static and time-invariant quantizer, defined as follows:

$$f(y) = [f_1(y_1), f_2(y_2), \ldots, f_q(y_q)]^T, \tag{8.6}$$

where y_i $(i \in \{1, 2, \ldots, q\})$ is the ith component of y and $f_i(-y_i) = -f_i(y_i)$. The logarithmic quantizer is characterized by the set of quantization levels as

$$\begin{aligned} U_i =& \{\pm u_i^j : u_i^j = \rho_i^j u_{i0}, j = \pm 1, \pm 2, \ldots\} \cup \{0\} \\ & (0 < \rho_i < 1,\ u_{i0} > 0). \end{aligned} \tag{8.7}$$

Each of quantization level u_i^j corresponds to a segment of the ith component of the output that the quantizer maps the whole segments to the quantization level as shown in Fig. 8.2. The parameter ρ_i is the quantization density. The associate quantizer $f_i(y_i)$ is defined as

$$f_i(y_i) = \begin{cases} u_i^j, & \text{if } \frac{1}{1+\delta_i} u_i^j \leq y_i \leq \frac{1}{1-\delta_i} u_i^j, \\ 0, & \text{if } y_i = 0, \\ -f_i(-y_i), & \text{if } y_i < 0, \end{cases} \tag{8.8}$$

where $\delta_i = \frac{1-\rho_i}{1+\rho_i}$.

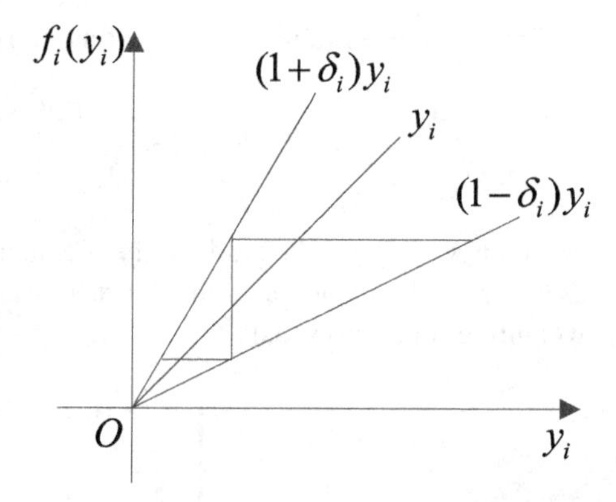

Fig. 8.2 Logarithmic quantizer

We adopt the sector bound approach developed in [1] to solve the quantization errors:

$$f(y_k) - y_k = \Delta_k y_k, \tag{8.9}$$

where $\Delta_k = diag\{\Delta_{1k}, \Delta_{2k}, \ldots, \Delta_{qk}\}$, and $|\Delta_{ik}| \leq \delta_i$ $(i = \{1, 2, ...q\})$. The signal received by the filter can be described as

$$y_{fk} = (I + \Delta_k)y_k. \tag{8.10}$$

8.2.3 Filtering Error Systems

The full-order filter is constructed as

$$\begin{cases} \hat{x}_{k+1} = A_{flh}\hat{x}_k + B_{flh}y_{fk}, \\ \hat{z}_k = E_{flh}\hat{x}_k, \end{cases} \tag{8.11}$$

where

$$A_{flh} = \sum_{i=1}^{r} h_i A_{fli}, \ B_{flh} = \sum_{i=1}^{r} h_i B_{fli}, \ E_{flh} = \sum_{i=1}^{r} h_i E_{fli}.$$

From (8.3) and (8.11), we define $e_k = z_k - \hat{z}_k$, $\xi_k = \left[x_k^{\mathrm{T}} \ \hat{x}_k^{\mathrm{T}}\right]^{\mathrm{T}}$, and have the filtering error system as follows:

$$\begin{cases} \xi_{k+1} = (\bar{A}_{lh} + \bar{B}_{1lh}\Delta_k'\bar{C}_{lh})\xi_k + (\bar{B}_{2lh} + \bar{B}_{1lh}\Delta_k'\bar{D}_{lh})w_k, \\ e_k = \bar{E}_{lh}\xi_k, \end{cases} \tag{8.12}$$

where

$$\bar{A}_{lh} = \begin{bmatrix} A_{lh} & 0 \\ B_{flh}C_{lh} & A_{flh} \end{bmatrix}, \ \bar{B}_{1lh} = \begin{bmatrix} 0 \\ B_{flh}\Delta \end{bmatrix},$$
$$\bar{B}_{2lh} = \begin{bmatrix} B_{lh} \\ B_{flh}D_{lh} \end{bmatrix}, \ \bar{C}_{lh} = \begin{bmatrix} C_{lh} & 0 \end{bmatrix},$$
$$\bar{D}_{lh} = D_{lh}, \ \bar{E}_{lh} = \begin{bmatrix} E_{lh} & -E_{flh} \end{bmatrix},$$
$$\Delta'_k = \Delta^{-1}\Delta_k, \ \Delta = diag\{\delta_1, \delta_2, \ldots, \delta_q\}.$$

We give the following definition and lemma which are useful in the sequel.

Definition 8.1 The filtering error system (8.12) is said to be stochastically stable for any initial condition when $w_k \equiv 0$, if the following condition holds:

$$E\left\{\sum_{k=0}^{\infty} \|\xi_k\|_2^2 \mid \xi_0\right\} < \infty. \tag{8.13}$$

Lemma 8.1 ([2]) *Given appropriately dimensioned matrices Γ_1, Γ_2, and Γ_3 with $\Gamma_1 = \Gamma_1^T$, then*

$$\Gamma_1 + \Gamma_3\Delta_k\Gamma_2 + \Gamma_2^T\Delta_k^T\Gamma_3^T < 0 \tag{8.14}$$

holds for all Δ_k satisfying $\Delta_k^T\Delta_k \leq I$ if and only if for some $\varepsilon > 0$,

$$\Gamma_1 + \varepsilon^{-1}\Gamma_3\Gamma_3^T + \varepsilon\Gamma_2^T\Gamma_2 < 0. \tag{8.15}$$

The main purpose of this chapter is that for the given system (8.1) and a prescribed scalar $\gamma > 0$, design two kind filters (H_∞ and L_2–L_∞ filtering) in the form of (8.11) such that the following two requirements are satisfied:
(1) The filtering error system (8.12) is stochastically stable in the case of $w_k \equiv 0$;
(2) The filtering error system (8.12) ensures a noise attenuation level γ in the H_∞ or L_2–L_∞ sense. More specifically, under the zero-initial condition for any nonzero $w_k \in l_2[0, \infty)$, e_k satisfies:
for H_∞ filtering problem,

$$E\left(\sqrt{\sum_{k=0}^{\infty} ||e_k||_2^2}\right) < \gamma\sqrt{\sum_{k=0}^{\infty} ||w_k||_2^2}, \tag{8.16}$$

and for L_2–L_∞ filtering problem,

$$\sup_k \sqrt{E[||e_k||_2^2]} < \gamma\sqrt{\sum_{k=0}^{\infty} ||w_k||_2^2}. \tag{8.17}$$

8.3 H_∞ Filtering Design

We firstly propose a sufficient condition to guarantee the stochastic stability of the filtering error system (8.12) with H_∞ performance. Then, the solution to filter gains in (8.11) is given.

Theorem 8.1 *For a given $\gamma > 0$, the filtering error system (8.12) is stochastically stable with the H_∞ performance γ, if there exist matrices $P_h > 0$, $P_{h^+} > 0$, $Q_{lh} > 0$ and a scalar $\varepsilon > 0$ for $h \in \rho$, $h^+ \triangleq (h_1(\theta_{k+1}), h_2(\theta_{k+1}), \cdots, h_r(\theta_{k+1})) \in \rho$, and $l \in \{1, 2, \ldots, L\}$ satisfying*

$$\sum_{l=1}^{L} \pi_l Q_{lh} < P_h, \tag{8.18}$$

$$\begin{bmatrix} -P_{h^+} & 0 & P_{h^+}\bar{A}_{lh} & P_{h^+}\bar{B}_{2lh} & P_{h^+}\bar{B}_{1lh} & 0 \\ * & -I & \bar{E}_{lh} & 0 & 0 & 0 \\ * & * & -Q_{lh} & 0 & 0 & \varepsilon\bar{C}_{lh}^T \\ * & * & * & -\gamma^2 I & 0 & \varepsilon\bar{D}_{lh}^T \\ * & * & * & * & -\varepsilon I & 0 \\ * & * & * & * & * & -\varepsilon I \end{bmatrix} < 0. \tag{8.19}$$

Proof Applying Schur Complement to (8.19) obtains

$$\begin{aligned} &\begin{bmatrix} -P_{h^+} & 0 & P_{h^+}\bar{A}_{lh} & P_{h^+}\bar{B}_{2lh} \\ * & -I & \bar{E}_{lh} & 0 \\ * & * & -Q_{lh} & 0 \\ * & * & * & -\gamma^2 I \end{bmatrix} + \varepsilon \begin{bmatrix} 0 \\ 0 \\ \bar{C}_{lh}^T \\ \bar{D}_{lh}^T \end{bmatrix} \begin{bmatrix} 0 \\ 0 \\ \bar{C}_{lh}^T \\ \bar{D}_{lh}^T \end{bmatrix}^T \\ &+ \varepsilon^{-1} \begin{bmatrix} P_{h^+}\bar{B}_{1lh} \\ 0 \\ 0 \\ 0 \end{bmatrix} \begin{bmatrix} P_{h^+}\bar{B}_{1lh} \\ 0 \\ 0 \\ 0 \end{bmatrix}^T < 0. \end{aligned} \tag{8.20}$$

Then due to $\Delta_k'^T \Delta_k' < I$, the following inequality holds from (8.20) by Lemma 8.1:

$$\begin{bmatrix} -P_{h^+} & 0 & P_{h^+}\kappa_{13} & P_{h^+}\kappa_{14} \\ * & -I & \bar{E}_{lh} & 0 \\ * & * & -Q_{lh} & 0 \\ * & * & * & -\gamma^2 I \end{bmatrix} < 0, \tag{8.21}$$

where

$$\kappa_{13} = \bar{A}_{lh} + \bar{B}_{1lh}\Delta_k'\bar{C}_{lh}, \quad \kappa_{14} = \bar{B}_{2lh} + \bar{B}_{1lh}\Delta_k'\bar{D}_{lh}.$$

By using Schur Complement to (8.21), we have

$$G_{1l} < Q_{lh}, \tag{8.22}$$

and

$$G_{2l} + \begin{bmatrix} \bar{E}_{lh}^T \\ 0 \end{bmatrix} \begin{bmatrix} \bar{E}_{lh}^T \\ 0 \end{bmatrix}^T < \begin{bmatrix} Q_{lh} & 0 \\ 0 & \gamma^2 I \end{bmatrix}, \tag{8.23}$$

where

$$G_{1l} = \kappa_{13}^T P_{h^+} \kappa_{13}, \; G_{2l} = \begin{bmatrix} \kappa_{13}^T \\ \kappa_{14}^T \end{bmatrix} P_{h^+} \begin{bmatrix} \kappa_{13} & \kappa_{14} \end{bmatrix}.$$

Because of $\sum_{l=1}^{L} \pi_l = 1$, we obtain the following inequalities according to (8.18):

$$\sum_{l=1}^{L} \pi_l G_{1l} < \sum_{l=1}^{L} \pi_l Q_{lh} < P_h, \tag{8.24}$$

and

$$\begin{aligned} \sum_{l=1}^{L} \pi_l \left(G_{2l} + \begin{bmatrix} \bar{E}_{lh}^T \\ 0 \end{bmatrix} \begin{bmatrix} \bar{E}_{lh}^T \\ 0 \end{bmatrix}^T \right) &< \begin{bmatrix} \sum_{l=1}^{L} \pi_l Q_{lh} & 0 \\ 0 & \gamma^2 I \end{bmatrix} \\ &< \begin{bmatrix} P_h & 0 \\ 0 & \gamma^2 I \end{bmatrix}. \end{aligned} \tag{8.25}$$

And it is easily observed that

$$G_{3l} = \sum_{l=1}^{L} \pi_l \left(G_{2l} + \begin{bmatrix} \bar{E}_{lh}^T \\ 0 \end{bmatrix} \begin{bmatrix} \bar{E}_{lh}^T \\ 0 \end{bmatrix}^T \right) - \begin{bmatrix} P_h & 0 \\ 0 & \gamma^2 I \end{bmatrix} < 0. \tag{8.26}$$

Choose the following Lyapunov function for the filtering error system (8.12):

$$V_k = \xi_k^T P_h \xi_k. \tag{8.27}$$

Define $\Delta V_k = E\{V_{k+1} | \xi_k\} - V_k$, where $V_{k+1} = \xi_{k+1}^T P_{h^+} \xi_{k+1}$. Considering $E\{\pi_{kl_1} \pi_{kl_2}\} = \begin{cases} \pi_{l_1}, & l_1 = l_2, \\ 0, & l_1 \neq l_2, \end{cases}$ when $w_k \equiv 0$ along the trajectory of system (8.12), we have

$$E\{\Delta V_k\} = \xi_k^T \left(\sum_{l=1}^{L} \pi_l G_{1l} - P_h \right) \xi_k < 0, \tag{8.28}$$

which guarantees the stochastic stability of the filtering error system (8.12).

Next, to establish H_∞ performance γ for the filtering error system (8.12), we assume the zero initial condition, and consider the following function:

$$J = E\{\Delta V_n\} + e_n^T e_n - \gamma^2 w_n^T w_n. \tag{8.29}$$

Taking mathematical expectation on both sides, we have

$$J = E\{V_{n+1}\} - E\{V_n\} + E\{e_n^{\mathrm{T}} e_n\} - \gamma^2 w_n^{\mathrm{T}} w_n \tag{8.30}$$

$$= \begin{bmatrix} \xi_n \\ w_n \end{bmatrix}^T G_{3l} \begin{bmatrix} \xi_n \\ w_n \end{bmatrix} < 0. \tag{8.31}$$

For $n = 0, 1, 2, ..., \infty$ summing up both sides, considering $E\{V_n\} \geq 0$ for all $n \geq 0$, under the zero initial condition, we have

$$E\{\sum_{n=0}^{\infty} e_n^{\mathrm{T}} e_n\} - \sum_{n=0}^{\infty} \gamma^2 w_n^{\mathrm{T}} w_n < 0. \tag{8.32}$$

Thus, it meets the H_∞ performance of the filtering error system. The proof is completed.

Now, it is time for us to design the filter of (8.11) based on Theorem 8.1 and the result is given as follows:

Theorem 8.2 *A filter in the form of (8.11) exists such that the filtering error system (8.12) is stochastically stable with H_∞ performance γ if there exist matrices*

$$\begin{bmatrix} P_{1i} & P_{2i} \\ * & P_{3i} \end{bmatrix} > 0, \begin{bmatrix} Q_{1li} & Q_{2li} \\ * & Q_{3li} \end{bmatrix} > 0, \begin{bmatrix} W_{1l} & W_{2l} \\ W_{3l} & W_{2l} \end{bmatrix},$$

$\hat{A}_{fli}$, $\hat{B}_{fli}$, $\hat{E}_{fli}$, *and a scalar* $\varepsilon > 0$ *for any* $l \in \{1, 2, \ldots, L\}$, $i, j, t \in \{1, 2, \ldots, r\}$ *satisfying*

$$\sum_{l=1}^{L} \pi_l \begin{bmatrix} Q_{1li} & Q_{2li} \\ * & Q_{3li} \end{bmatrix} < \begin{bmatrix} P_{1i} & P_{2i} \\ * & P_{3i} \end{bmatrix}, \tag{8.33}$$

$$\Theta_{lijt} = \begin{bmatrix} \theta_{1lt} & \theta_{2lt} & 0 & \theta_{3lij} & \hat{A}_{fli} & \theta_{4lij} & \hat{B}_{fli}\Delta & 0 \\ * & \theta_{5lt} & 0 & \theta_{6lij} & \hat{A}_{fli} & \theta_{7lij} & \hat{B}_{fli}\Delta & 0 \\ * & * & -I & E_{li} & -\hat{E}_{fli} & 0 & 0 & 0 \\ * & * & * & -Q_{1li} & -Q_{2li} & 0 & 0 & \varepsilon C_{li}^T \\ * & * & * & * & -Q_{3li} & 0 & 0 & 0 \\ * & * & * & * & * & -\gamma^2 I & 0 & \varepsilon D_{li}^T \\ * & * & * & * & * & * & -\varepsilon I & 0 \\ * & * & * & * & * & * & * & -\varepsilon I \end{bmatrix} < 0, \tag{8.34}$$

where

$$\theta_{1lt} = P_{1t} - W_{1l} - W_{1l}^T,\ \theta_{2lt} = P_{2t} - W_{2l} - W_{3l}^T,$$
$$\theta_{3lij} = W_{1l}A_{li} + \hat{B}_{fli}C_{lj},\ \theta_{4lij} = W_{1l}B_{li} + \hat{B}_{fli}D_{lj},$$

$$\theta_{5lt} = P_{3t} - W_{2l} - W_{2l}^T,\ \theta_{6lij} = W_{3l}A_{li} + \hat{B}_{fli}C_{lj},$$
$$\theta_{7lij} = W_{3l}B_{li} + \hat{B}_{fli}D_{lj}.$$

Furthermore, the parameters of the filter in (8.11) can be given by

$$A_{fli} = W_{2l}^{-1}\hat{A}_{fli},\ B_{fli} = W_{2l}^{-1}\hat{B}_{fli},\ E_{fli} = \hat{E}_{fli}. \tag{8.35}$$

Proof Firstly, suppose that there exist matrices $P_h, P_{h^+}, Q_{lh}, W_l, \hat{A}_{flh}, \hat{B}_{flh}, \hat{E}_{flh}$ satisfying (8.33) and (8.34). We utilize these matrices to define the following functions:

$$P_h = \begin{bmatrix} P_{1h} & P_{2h} \\ * & P_{3h} \end{bmatrix} = \sum_{i=1}^{r} h_i \begin{bmatrix} P_{1i} & P_{2i} \\ * & P_{3i} \end{bmatrix},$$
$$P_{h^+} = \begin{bmatrix} P_{1h^+} & P_{2h^+} \\ * & P_{3h^+} \end{bmatrix} = \sum_{t=1}^{r} h_t^+ \begin{bmatrix} P_{1t} & P_{2t} \\ * & P_{3t} \end{bmatrix},$$
$$Q_{lh} = \begin{bmatrix} Q_{1lh} & Q_{2lh} \\ * & Q_{3lh} \end{bmatrix} = \sum_{i=1}^{r} h_i \begin{bmatrix} Q_{1li} & Q_{2li} \\ * & Q_{3li} \end{bmatrix},\ W_l = \begin{bmatrix} W_{1l} & W_{2l} \\ W_{3l} & W_{2l} \end{bmatrix},$$
$$\hat{A}_{flh} = W_{2l}A_{flh} = \sum_{i=1}^{r} h_i W_{2l}A_{fli} = \sum_{i=1}^{r} h_i \hat{A}_{fli},$$
$$\hat{B}_{flh} = W_{2l}B_{flh} = \sum_{i=1}^{r} h_i W_{2l}B_{fli} = \sum_{i=1}^{r} h_i \hat{B}_{fli},$$
$$\hat{E}_{flh} = E_{flh} = \sum_{i=1}^{r} h_i E_{fli} = \sum_{i=1}^{r} h_i \hat{E}_{fli}.$$

Thus, we obtain

$$\sum_{i=1}^{r}\sum_{j=1}^{r}\sum_{t=1}^{r} h_i h_j h_t^+ \Theta_{lijt} = \begin{bmatrix} \theta_1 & \theta_2 & 0 & \theta_3 & \hat{A}_{flh} & \theta_4 & \theta_8 & 0 \\ * & \theta_5 & 0 & \theta_6 & \hat{A}_{flh} & \theta_7 & \theta_8 & 0 \\ * & * & -I & E_{lh} & -\hat{E}_{flh} & 0 & 0 & 0 \\ * & * & * & -Q_{1lh} & -Q_{2lh} & 0 & 0 & \varepsilon C_{lh}^T \\ * & * & * & * & -Q_{3lh} & 0 & 0 & 0 \\ * & * & * & * & * & -\gamma^2 I & 0 & \varepsilon D_{lh}^T \\ * & * & * & * & * & * & -\varepsilon I & 0 \\ * & * & * & * & * & * & * & -\varepsilon I \end{bmatrix}$$
$$= \begin{bmatrix} \lambda & 0 & W_l\bar{A}_{lh} & W_l\bar{B}_{2lh} & W_l\bar{B}_{1lh} & 0 \\ * & -I & \bar{E}_{lh} & 0 & 0 & 0 \\ * & * & -Q_{lh} & 0 & 0 & \varepsilon\bar{C}_{lh}^T \\ * & * & * & -\gamma^2 I & 0 & \varepsilon\bar{D}_{lh}^T \\ * & * & * & * & -\varepsilon I & 0 \\ * & * & * & * & * & -\varepsilon I \end{bmatrix} < 0, \tag{8.36}$$

where

$$\theta_1 = P_{1h^+} - W_{1l} - W_{1l}^T,\ \theta_2 = P_{2h^+} - W_{2l} - W_{3l}^T,$$
$$\theta_3 = W_{1l}A_{lh} + \hat{B}_{flh}C_{lh},\ \theta_4 = W_{1l}B_{lh} + \hat{B}_{flh}D_{lh},$$
$$\theta_5 = P_{3h^+} - W_{2l} - W_{2l}^T,\ \theta_6 = W_{3l}A_{lh} + \hat{B}_{flh}C_{lh},$$
$$\theta_7 = W_{3l}B_{lh} + \hat{B}_{flh}D_{lh},\ \theta_8 = \hat{B}_{flh}\Delta,\ \lambda = P_{h^+} - W_l - W_l^T.$$

Noting that $P_{h^+} > 0$, we have

$$P_{h^+} - W_l - W_l^T > -W_l P_{h^+}^{-1} W_l^T. \tag{8.37}$$

Thus it follows from (8.36) that

$$\begin{bmatrix} -W_l P_{h^+}^{-1} W_l^T & 0 & W_l\bar{A}_{lh} & W_l\bar{B}_{2lh} & W_l\bar{B}_{1lh} & 0 \\ * & -I & \bar{E}_{lh} & 0 & 0 & 0 \\ * & * & -Q_{lh} & 0 & 0 & \varepsilon\bar{C}_{lh}^T \\ * & * & * & -\gamma^2 I & 0 & \varepsilon\bar{D}_{lh}^T \\ * & * & * & * & -\varepsilon I & 0 \\ * & * & * & * & * & -\varepsilon I \end{bmatrix} < 0. \tag{8.38}$$

By pre-multiplying $\mathrm{diag}\{W_l^{-1}, I, I, I, I, I\}$ and post-multiplying $\mathrm{diag}\{W_l^{-T}, I, I, I, I, I\}$ to (8.38), we obtain

$$\begin{bmatrix} P_{h^+}^{-1} & 0 & \bar{A}_{lh} & \bar{B}_{2lh} & \bar{B}_{1lh} & 0 \\ * & -I & \bar{E}_{lh} & 0 & 0 & 0 \\ * & * & -Q_{lh} & 0 & 0 & \varepsilon\bar{C}_{lh}^T \\ * & * & * & -\gamma^2 I & 0 & \varepsilon\bar{D}_{lh}^T \\ * & * & * & * & -\varepsilon I & 0 \\ * & * & * & * & * & -\varepsilon I \end{bmatrix} < 0. \tag{8.39}$$

By pre-multiplying and post-multiplying $\mathrm{diag}\{P_{h^+}^{-1}, I, I, I, I, I\}$ to (8.19), we have (8.39).

Thus, when (8.34) holds, (8.19) holds. Moreover, (8.33) is equivalent to (8.18). From above analysis, we obtain

$$\hat{A}_{fli} = W_{2l}A_{fli},\ \hat{B}_{fli} = W_{2l}B_{fli},\ \hat{E}_{fli} = E_{fli}. \tag{8.40}$$

Thus,

$$A_{fli} = W_{2l}^{-1}\hat{A}_{fli},\ B_{fli} = W_{2l}^{-1}\hat{B}_{fli},\ E_{fli} = \hat{E}_{fli}. \tag{8.41}$$

The proof is completed.

Remark 8.3 According to Theorem 8.2, the feasibility of LMIs (8.33) and (8.34) provides a solution to the H_∞ filtering problem. The optimal H_∞ performance γ^* can be obtained via dealing with the optimal problem as follows:

$$\min \quad \sigma \quad \text{subject to (8.33) and (8.34) with } \sigma = \gamma^2. \tag{8.42}$$

8.4 L_2–L_∞ Filtering Design

In this section, a sufficient condition for the stochastic stability of the filtering error system (8.12) with L_2–L_∞ performance is proposed and then the solution to filter gains in (8.11) is developed. Unless otherwise defined, we associate the same meaning to the notations used in the H_∞ filtering case.

Theorem 8.4 *For a given $\gamma > 0$, the filtering error system (8.12) is stochastically stable with L_2–L_∞ performance γ, if there exist matrices $P_h > 0$, $P_{h^+} > 0$, $Q_{lh} > 0$ and a scalar $\varepsilon > 0$ for $h \in \rho$, $h^+ \triangleq (h_1(\theta_{k+1}), h_2(\theta_{k+1}), \cdots, h_r(\theta_{k+1})) \in \rho$, and $l \in \{1, 2, \ldots, L\}$ satisfying*

$$\sum_{l=1}^{L} \pi_l Q_{lh} < P_h, \tag{8.43}$$

$$\begin{bmatrix} -P_{h^+} & P_{h^+}\bar{A}_{lh} & P_{h^+}\bar{B}_{2lh} & P_{h^+}\bar{B}_{1lh} & 0 \\ * & -Q_{lh} & 0 & 0 & \varepsilon\bar{C}_{lh}^T \\ * & * & -I & 0 & \varepsilon\bar{D}_{lh}^T \\ * & * & * & -\varepsilon I & 0 \\ * & * & * & * & -\varepsilon I \end{bmatrix} < 0, \tag{8.44}$$

$$\begin{bmatrix} -P_h & -\bar{E}_{lh}^T \\ * & -\gamma^2 I \end{bmatrix} \leq 0. \tag{8.45}$$

Proof By following the same line of the stability proof of Theorem 8.1, we can have that the filtering error system (8.12) with $w_k = 0$ is stochastically stable, and the proof is omitted here.

Now, to establish the L_2–L_∞ performance γ for the filtering error system (8.12), we assume the zero initial condition and consider the following function:

$$\begin{aligned} J &= \sum_{n=0}^{k-1} E\{\Delta V_n - w_n^T w_n\} \\ &= \sum_{n=0}^{k-1} E\left\{\begin{bmatrix} \xi_n \\ w_n \end{bmatrix}^T \left(\sum_{l=1}^{L} \pi_l G_{2l} - \begin{bmatrix} P_h & 0 \\ 0 & I \end{bmatrix}\right) \begin{bmatrix} \xi_n \\ w_n \end{bmatrix}\right\}. \end{aligned} \tag{8.46}$$

Note that we use the same Lyapunov function in Theorem 8.1.

Applying Schur Complement to (8.44), we obtain

$$\begin{aligned}&\begin{bmatrix}-P_{h^+} & P_{h^+}\bar{A}_{lh} & P_{h^+}\bar{B}_{2lh}\\ * & -Q_{lh} & 0\\ * & * & -I\end{bmatrix}+\varepsilon\begin{bmatrix}0\\ \bar{C}_{lh}^T\\ \bar{D}_{lh}^T\end{bmatrix}\begin{bmatrix}0\\ \bar{C}_{lh}^T\\ \bar{D}_{lh}^T\end{bmatrix}^T\\ &+\varepsilon^{-1}\begin{bmatrix}P_{h^+}\bar{B}_{1lh}\\ 0\\ 0\end{bmatrix}\begin{bmatrix}P_{h^+}\bar{B}_{1lh}\\ 0\\ 0\end{bmatrix}^T<0.\end{aligned} \tag{8.47}$$

Then by using Lemma 8.1, it follows from (8.47) that

$$\begin{bmatrix}-P_{h^+} & P_{h^+}\kappa_{12} & P_{h^+}\kappa_{13}\\ * & -Q_{lh} & 0\\ * & * & -I\end{bmatrix}<0. \tag{8.48}$$

Via adopting Schur Complement, we have

$$G_{2l}<\begin{bmatrix}Q_{lh} & 0\\ 0 & I\end{bmatrix}. \tag{8.49}$$

It follows from (8.45) that

$$\bar{E}_{lh}^T\bar{E}_{lh}<\gamma^2 P_h. \tag{8.50}$$

We obtain the following inequality from (8.43):

$$\sum_{l=1}^{L}\pi_l G_{2l}<\begin{bmatrix}\sum_{l=1}^{L}\pi_l Q_{lh} & 0\\ 0 & I\end{bmatrix}<\begin{bmatrix}P_h & 0\\ 0 & I\end{bmatrix}. \tag{8.51}$$

It implies that

$$J=\sum_{n=0}^{k-1}E\{\Delta V_n-w_n^T w_n\}<0. \tag{8.52}$$

Under the zero initial condition, we have

$$E\{V_k\}=E\{\xi_k^T P_h\xi_k\}<\sum_{n=0}^{k-1}w_n^T w_n. \tag{8.53}$$

On the other hand, we obtain the following inequality from (8.50):

$$\begin{aligned}E\{e_k^T e_k\}&=E\{\xi_k^T\bar{E}_{lh}^T\bar{E}_{lh}\xi_k\}<E\{\xi_k^T\gamma^2 P_h\xi_k\}\\ &<\gamma^2\sum_{n=0}^{k-1}w_n^T w_n<\gamma^2\sum_{n=0}^{\infty}w_n^T w_n,\end{aligned} \tag{8.54}$$

which implies (8.17) holds under the zero initial condition for any nonzero $w_k \in l_2(0, \infty]$. The proof is completed.

Now, we are ready to design the filter in the form of (8.11) based on Theorem 8.4 and the result is given as follows:

Theorem 8.5 *A filter in the form of (8.11) exists such that the filtering error system (8.12) is stochastically stable with L_2–L_∞ performance γ if there exist matrices*

$$\begin{bmatrix} P_{1i} & P_{2i} \\ * & P_{3i} \end{bmatrix} > 0, \begin{bmatrix} Q_{1li} & Q_{2li} \\ * & Q_{3li} \end{bmatrix} > 0, \begin{bmatrix} W_{1l} & W_{2l} \\ W_{3l} & W_{2l} \end{bmatrix},$$

$\hat{A}_{fli}$, $\hat{B}_{fli}$, $\hat{E}_{fli}$, *and a scalar* $\varepsilon > 0$ *for any* $l \in \{1, 2, \ldots, L\}$, $i, j, t \in \{1, 2, \ldots, r\}$ *satisfying*

$$\sum_{l=1}^{L} \pi_l \begin{bmatrix} Q_{1li} & Q_{2li} \\ * & Q_{3li} \end{bmatrix} < \begin{bmatrix} P_{1i} & P_{2i} \\ * & P_{3i} \end{bmatrix}, \tag{8.55}$$

$$\Theta_{lijt} = \begin{bmatrix} \theta_{1lt} & \theta_{2lt} & \theta_{3lij} & \hat{A}_{fli} & \theta_{4lij} & \hat{B}_{fli}\Delta & 0 \\ * & \theta_{5lt} & \theta_{6lij} & \hat{A}_{fli} & \theta_{7lij} & \hat{B}_{fli}\Delta & 0 \\ * & * & -Q_{1li} & -Q_{2li} & 0 & 0 & \varepsilon C_{li}^T \\ * & * & * & -Q_{3li} & 0 & 0 & 0 \\ * & * & * & * & -I & 0 & \varepsilon D_{li}^T \\ * & * & * & * & * & -\varepsilon I & 0 \\ * & * & * & * & * & * & -\varepsilon I \end{bmatrix} < 0, \tag{8.56}$$

$$\begin{bmatrix} -P_{1i} & -P_{2i} & -E_{li}^T \\ * & -P_{3i} & \hat{E}_{fli}^T \\ * & * & -\gamma^2 I \end{bmatrix} \leq 0. \tag{8.57}$$

Furthermore, the parameters of the filter in (8.11) can be given by

$$A_{fli} = W_{2l}^{-1}\hat{A}_{fli}, \ B_{fli} = W_{2l}^{-1}\hat{B}_{fli}, \ E_{fli} = \hat{E}_{fli}. \tag{8.58}$$

Proof The proof procedures follow similarly from Theorem 8.2, and hence the details are omitted here.

Remark 8.6 As in the case of H_∞ filtering, the feasibility of LMIs (8.55), (8.56) and (8.57) in Theorem 8.5 provides a solution to the L_2–L_∞ filtering problem. The optimal L_2–L_∞ performance γ^* can be obtained through solving the following optimal problem:

$$\min \quad \sigma \quad \text{subject to (8.55), (8.56) and (8.57) with } \sigma = \gamma^2. \tag{8.59}$$

Remark 8.7 It is worth noting that the number of scalar decision variables in LMIs of Theorem 8.2 is the same as that of Theorem 8.5, which is $(3L + 2r + 3Lr)n^2 +$

$(1 + L + qL + vL)nr$. On the other hand, there are $rL + r^3L$ and $2rL + r^3L$ LMIs for Theorems 8.2 and 8.5, respectively. And it implies the L_2–L_∞ filtering has much more computational burden than the H_∞ filtering.

8.5 Illustrative Example

In this section, we present an example to show the effectiveness of the proposed filtering design approaches. Consider the following model:

$$\begin{cases} x_{k+1} = A_{\mathcal{T}_k i} x_k + B_{\mathcal{T}_k i} w_k, \\ y_k = C_{\mathcal{T}_k i} x_k + D_{\mathcal{T}_k i} w_k, \\ z_k = E_{\mathcal{T}_k i} x_k, \end{cases}$$

Mode 1:

$$\begin{aligned} A_{11} &= \begin{bmatrix} 0.60 & 0.41 \\ -0.11 & 0.43 \end{bmatrix}, \; A_{12} = \begin{bmatrix} 0.80 & 0.08 \\ -0.19 & 0.61 \end{bmatrix}, \\ B_{11} &= \begin{bmatrix} 0.11 \\ 0.19 \end{bmatrix}, \; B_{12} = \begin{bmatrix} 0.21 \\ 0.29 \end{bmatrix}, \; C_{11} = C_{12} = \begin{bmatrix} 1 & 0 \end{bmatrix}, \\ D_{11} &= D_{12} = 1, \, E_{11} = E_{12} = \begin{bmatrix} 1 & 0 \end{bmatrix}; \end{aligned}$$

Mode 2:

$$\begin{aligned} A_{21} &= \begin{bmatrix} 0.60 & 0.51 \\ -0.18 & 0.42 \end{bmatrix}, \; A_{22} = \begin{bmatrix} 0.91 & 0.09 \\ -0.13 & 0.70 \end{bmatrix}, \\ B_{21} &= \begin{bmatrix} 0.30 \\ 0.20 \end{bmatrix}, \; B_{22} = \begin{bmatrix} 0.10 \\ 0.20 \end{bmatrix}, \; C_{21} = C_{22} = \begin{bmatrix} 1 & 0 \end{bmatrix}, \\ D_{21} &= D_{22} = 1, \; E_{21} = E_{22} = \begin{bmatrix} 1 & 0 \end{bmatrix}. \end{aligned}$$

To show effectiveness of the obtained results, we adopt the following normalized fuzzy weighting functions:

$$h_1 = 0.5(\sin^2(x_{1k}) + \sin^2(x_{2k})), \; h_2 = 1 - h_1.$$

The switching probabilities are $\pi_1 = 0.3, \;\; \pi_2 = 0.7$, and the quantization density ρ is assumed to be 0.8182.

By using Matlab Toolbox to solve LMIs in Theorem 8.2, the optimal H_∞ performance γ^* is 0.1663, and the corresponding parameters of the mode-dependent filter are:

Mode-dependent filter 1:

$$A_{f11} = \begin{bmatrix} 0.4786 & 0.4064 \\ -0.2749 & 0.4300 \end{bmatrix}, \; A_{f12} = \begin{bmatrix} 0.5769 & 0.0757 \\ -0.4555 & 0.6158 \end{bmatrix},$$
$$B_{f11} = \begin{bmatrix} -0.1178 \\ -0.1787 \end{bmatrix}, \; B_{f12} = \begin{bmatrix} -0.2164 \\ -0.2791 \end{bmatrix},$$
$$E_{f11} = \begin{bmatrix} -0.9960 & -0.0028 \end{bmatrix}, \; E_{f12} = \begin{bmatrix} -0.9999 & -0.0008 \end{bmatrix};$$

Mode-dependent filter 2:

$$A_{f21} = \begin{bmatrix} 0.3009 & 0.5077 \\ -0.3533 & 0.4292 \end{bmatrix}, \; A_{f22} = \begin{bmatrix} 0.7970 & 0.0875 \\ -0.3191 & 0.7001 \end{bmatrix},$$
$$B_{f21} = \begin{bmatrix} -0.3006 \\ -0.1891 \end{bmatrix}, \; B_{f22} = \begin{bmatrix} -0.1081 \\ -0.1936 \end{bmatrix},$$
$$E_{f21} = \begin{bmatrix} -0.9960 & -0.0028 \end{bmatrix}, \; E_{f22} = \begin{bmatrix} -0.9999 & -0.0008 \end{bmatrix}.$$

Via solving LMIs in Theorem 8.5, we have that the optimal L_2–L_∞ performance γ^* is 0.0837, and the corresponding parameters of the mode-dependent filter are:
Mode-dependent filter 1:

$$A_{f11} = \begin{bmatrix} 0.4830 & 0.4092 \\ -0.2900 & 0.4311 \end{bmatrix}, \; A_{f12} = \begin{bmatrix} 0.5856 & 0.0782 \\ -0.4635 & 0.6148 \end{bmatrix},$$
$$B_{f11} = \begin{bmatrix} -0.1146 \\ -0.1845 \end{bmatrix}, \; B_{f12} = \begin{bmatrix} -0.2121 \\ -0.2816 \end{bmatrix},$$
$$E_{f11} = \begin{bmatrix} -0.9847 & 0.0069 \end{bmatrix}, \; E_{f12} = \begin{bmatrix} -0.9840 & 0.0070 \end{bmatrix};$$

Mode-dependent filter 2:

$$A_{f21} = \begin{bmatrix} 0.3058 & 0.5136 \\ -0.3677 & 0.4264 \end{bmatrix}, \; A_{f22} = \begin{bmatrix} 0.8047 & 0.0882 \\ -0.3210 & 0.7005 \end{bmatrix},$$
$$B_{f21} = \begin{bmatrix} -0.2973 \\ -0.1947 \end{bmatrix}, \; B_{f22} = \begin{bmatrix} -0.1034 \\ -0.1947 \end{bmatrix},$$
$$E_{f21} = \begin{bmatrix} -0.9847 & 0.0069 \end{bmatrix}, \; E_{f22} = \begin{bmatrix} -0.9840 & 0.0070 \end{bmatrix}.$$

To verify the performance of the designed filter, we assume that the external disturbance w_k is

$$w_k = \begin{cases} 1 < w_k < 2, & 1 \le k \le 10, \\ -2 < w_k < -1, & 11 \le k \le 20, \\ 0, & \text{elsewhere.} \end{cases}$$

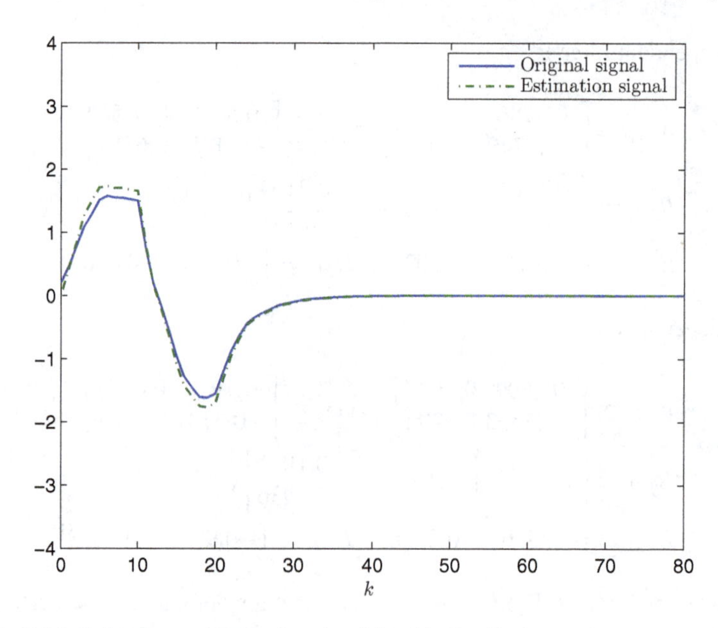

Fig. 8.3 Original signal z_k and estimation signal $\hat{z}_k$ with H_∞ filtering performance

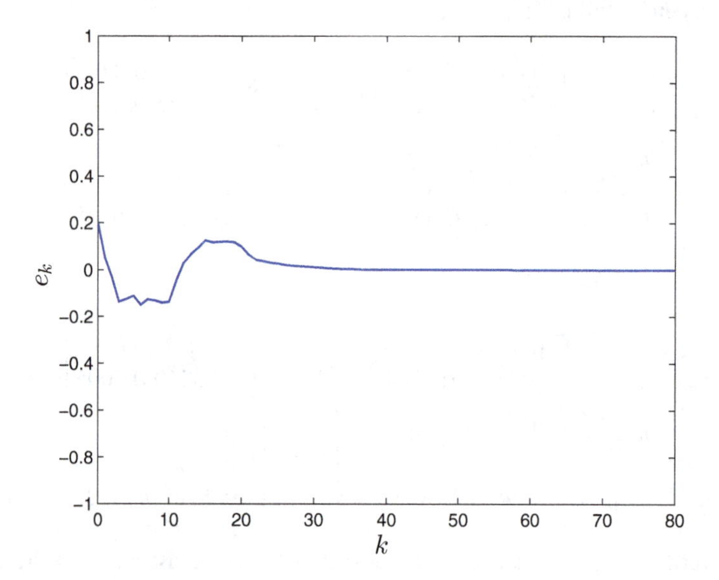

Fig. 8.4 Estimation error with H_∞ filtering performance

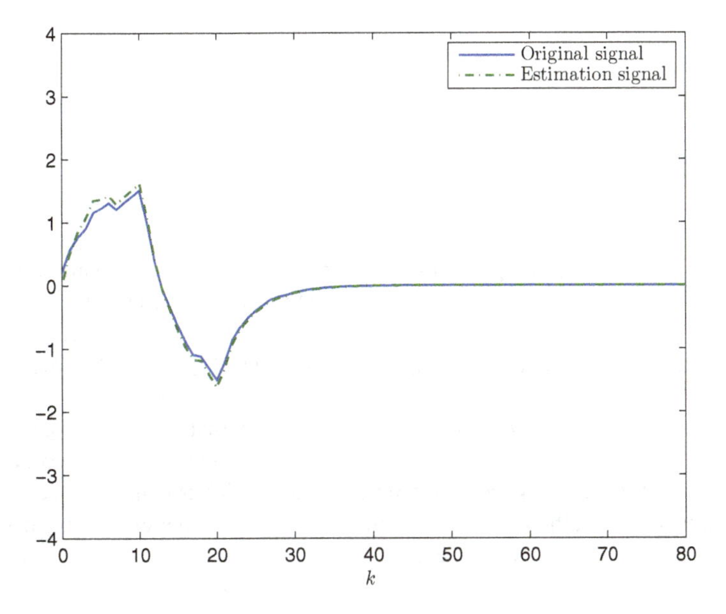

Fig. 8.5 Original signal z_k and estimation signal $\hat{z}_k$ with L_2–L_∞ filtering performance

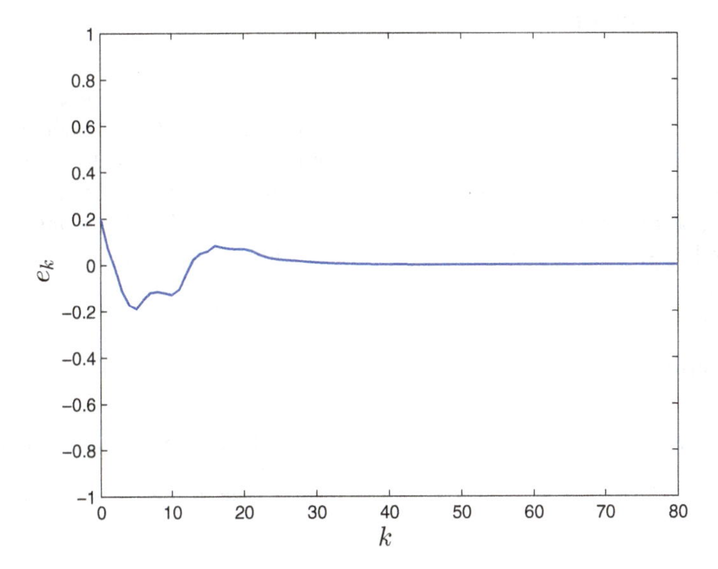

Fig. 8.6 Estimation error with L_2–L_∞ filtering performance

Under initial conditions $x_0 = \begin{bmatrix} 0.2 & -0.2 \end{bmatrix}^{\mathrm{T}}$ and $\hat{x}_0 = \begin{bmatrix} 0 & 0 \end{bmatrix}^{\mathrm{T}}$, Figs. 8.3 and 8.4 show the results of z_k, $\hat{z}_k$, and the estimation error under the external disturbance of the H_∞ filtering, respectively. Figures 8.5 and 8.6 present the corresponding results of the L_2–L_∞ filtering. These responses demonstrate that the proposed approaches are effective.

Table 8.1 Optimal performance according to different ρ

ρ	0.1	0.3	0.5	0.7	0.9
γ^* of H_∞ filter	1.0534	0.7260	0.4890	0.2829	0.0888
γ^* of L_2–L_∞ filter	0.4234	0.3424	0.2464	0.1433	0.0445

Furthermore, by changing the quantization density ρ, we can obtain the optimal H_∞ or L_2–L_∞ performance γ^* corresponding to different ρ as shown in Table 8.1. It is clear that the H_∞ (or L_2–L_∞) performance γ^* decreases as ρ increases. In other words, when denser quantizer is employed, the better H_∞ or L_2–L_∞ performance can be achieved.

Meanwhile, from Table 8.1, we can observe clearly that γ^* of L_2–L_∞ filtering is smaller than that of H_∞ filtering in this chapter. Since they are two different kind of characteristic filters, it is difficult and unfair to say which is better. Hence, if we are more concerned with the peak value of the output than output energy, we had better apply the L_2–L_∞ filtering approach.

8.6 Conclusion

In this chapter, we have investigated the problems of H_∞ and L_2–L_∞ filtering for discrete-time systems based on the T–S fuzzy model. Switching phenomena and quantization effects are taken into consideration simultaneously. Using the fuzzy-basis-dependent Lyapunov function ensures the stability and a given H_∞ or L_2–L_∞ filtering performance index of the filtering error system. The filter gains have been obtained by solving the corresponding convex optimization problem.

References

1. Fu, M., Xie, L.: The sector bound approach to quantized feedback control. IEEE Trans. Autom. Control **50**(11), 1698–1711 (2005)
2. Zhang, C., Feng, G., Gao, H., Qiu, J.: H_∞ filtering for nonlinear discrete-time systems subject to quantization and packet dropouts. IEEE Trans. Fuzzy Syst. **19**(2), 353–365 (2011)

Chapter 9
Reliable Filter Design of Fuzzy Switched Systems with Imprecise Modes

9.1 Introduction

The problem of sensor failures is inevitable in real-world control systems due to harsh working environment, power supply instability, inescapable component aging and so on [1–3]. Much more attention has been paid to designing a reliable filter that can tolerate the admissible failures and work successfully, such as the reliable filter design for T–S fuzzy systems [4] and the adaptive reliable filtering problem for continuous-time linear systems [5].

In this chapter, the asynchronous and reliable L_2–L_∞ filter design problem is investigated for discrete-time nonlinear MJSs, which are modeled by the T–S fuzzy model. The modes of the designed filter are observed by obeying some transition probabilities described by the nonstationary Markov jump model, which are imprecise to some extent. It results that the designed filter works asynchronously with the studied system. Via the Lyapunov function approach and the relaxation matrix technique, two approaches are proposed such that the filtering error system is stochastically stable with a prescribed L_2–L_∞ performance index.

9.2 Preliminary Analysis

The chapter studies the asynchronous filtering problem for T–S fuzzy MJSs with sensor failures, shown in Fig. 9.1. Now, consider the following T–S fuzzy systems with Markov jump parameters:

Plant rule i^{ρ_k}: IF θ_{1k} is $M_{i1}^{\rho_k}$, ..., and θ_{uk} is $M_{iu}^{\rho_k}$, THEN

$$\begin{cases} x_{k+1} = A_{\rho_k i} x_k + B_{\rho_k i} w_k, \\ y_k = C_{\rho_k i} x_k + D_{\rho_k i} w_k, \\ z_k = E_{\rho_k i} x_k, \end{cases} \tag{9.1}$$

S. Dong et al., *Control and Filtering of Fuzzy Systems with Switched Parameters*, Studies in Systems, Decision and Control 268,
https://doi.org/10.1007/978-3-030-35566-1_9

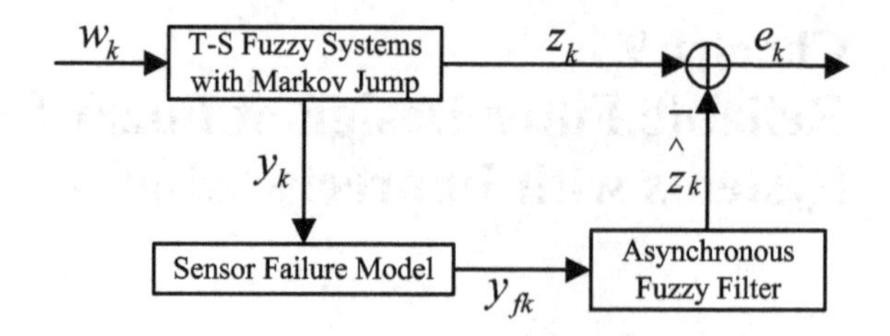

Fig. 9.1 Asynchronous filtering design with sensor failures

where $x_k \in R^n$ is the state vector; $y_k \in R^m$ is the measured output; $z_k \in R^q$ is the signal to be estimated; $w_k \in R^p$ is the external disturbance that belongs to $l_2[0, \infty)$; $(A_{\rho_k i}, B_{\rho_k i}, C_{\rho_k i}, D_{\rho_k i}, E_{\rho_k i})$ are known matrices with appropriate dimensions; $M_{ij}^{\rho_k}$, r and θ_{jk} $(i \in \mathsf{R} = \{1, 2, \ldots, r\},\ j = \{1, 2, \ldots, u\})$ are the fuzzy set, the number of fuzzy rules, and the premise variable, respectively. The variable ρ_k represents the discrete-time Markov chain, which takes values in $\mathsf{L} = \{1, 2, \ldots, L\}$ at each instant and L is the total number of jump modes with

$$\Pr\{\rho_{k+1} = b | \rho_k = a\} = \pi_{ab},\ a,\ b \in \mathsf{L}, \tag{9.2}$$

where π_{ab} $(0 \leq \pi_{ab} \leq 1)$ is the transition probability from mode a at time k to mode b at the next instant. The transition probability matrix is further defined as $\Pi = [\pi_{ab}]$. It is necessary to point out that $\sum_{b=1}^{L} \pi_{ab} = 1$.

When $\rho_k = a$, the overall fuzzy model is obtained by the T–S fuzzy inference method, as follows:

$$\begin{cases} x_{k+1} = A_{ah} x_k + B_{ah} w_k, \\ y_k = C_{ah} x_k + D_{ah} w_k, \\ z_k = E_{ah} x_k, \end{cases} \tag{9.3}$$

where

$$A_{ah} = \sum_{i=1}^{r} h_{ai}(\theta_k) A_{ai},\ B_{ah} = \sum_{i=1}^{r} h_{ai}(\theta_k) B_{ai},$$
$$C_{ah} = \sum_{i=1}^{r} h_{ai}(\theta_k) C_{ai},\ D_{ah} = \sum_{i=1}^{r} h_{ai}(\theta_k) D_{ai},$$
$$E_{ah} = \sum_{i=1}^{r} h_{ai}(\theta_k) E_{ai},\ h_{ai}(\theta_k) = \frac{\prod_{j=1}^{u} M_{ij}^{a}(\theta_{jk})}{\sum_{i=1}^{r} \prod_{j=1}^{u} M_{ij}^{a}(\theta_{jk})},$$

and it is assumed that $\prod_{j=1}^{u} M_{ij}^{a}(\theta_{jk}) \geq 0$. $M_{ij}^{a}(\theta_{jk})$ stands for the grade of the membership of θ_{jk} in M_{ij}^{a}. $h_{ai}(\theta_k)$ is the normalized fuzzy weighting function. We can easily conclude that $\sum_{i=1}^{r} h_{ai}(\theta_k) = 1$ and $h_{ai}(\theta_k) \geq 0$ for $i \in \mathsf{R}$.

In this chapter, it is assumed that sensor failures happen during transmitting signals, satisfying

$$y_{fkg} = s_g y_{kg},\ 0 \leq \underline{s}_g \leq s_g \leq \overline{s}_g \leq 1, \tag{9.4}$$

where y_{kg} and y_{fkg} ($g = \{1, 2, \ldots, m\}$) are the input and output signals of the gth sensor, respectively. m is the total number of sensors. s_g is unknown and denotes the gth sensor failure level. $\underline{s}_g$ and $\overline{s}_g$ are the given lower and upper bounds of s_g. It is worth noting that if $\underline{s}_g = \overline{s}_g = 0$, the outage of gth sensor will occur. $0 < \underline{s}_g < \overline{s}_g < 1$ means partial failure in the gth sensor. And when $\underline{s}_g = \overline{s}_g = 1$, there is no failure in the gth sensor.

Hence, the overall failure model is rewritten as

$$y_{fk} = S y_k, \tag{9.5}$$

where $S = \mathrm{diag}\{s_1, s_2, \ldots, s_m\}$.

We define

$$\begin{aligned}\acute{S} &= \mathrm{diag}\left\{\frac{\underline{s}_1 + \overline{s}_1}{2}, \frac{\underline{s}_2 + \overline{s}_2}{2}, \ldots, \frac{\underline{s}_m + \overline{s}_m}{2}\right\},\\ \grave{S} &= \mathrm{diag}\left\{\frac{\overline{s}_1 - \underline{s}_1}{2}, \frac{\overline{s}_2 - \underline{s}_2}{2}, \ldots, \frac{\overline{s}_m - \underline{s}_m}{2}\right\},\end{aligned}$$

and S is rewritten as

$$S = \acute{S} + \Lambda, \tag{9.6}$$

where $\Lambda = \mathrm{diag}\{\lambda_1, \lambda_2, \ldots, \lambda_m\}$ and $|\lambda_g| \leq (\overline{s}_g - \underline{s}_g)/2$.

Similar to the fuzzy controller design through the PDC approach, an asynchronous fuzzy filter is constructed as

$$\begin{cases} \hat{x}_{k+1} = A_{f\sigma_k h}\hat{x}_k + B_{f\sigma_k h} y_{fk}, \\ \hat{z}_k = E_{f\sigma_k h}\hat{x}_k, \end{cases} \tag{9.7}$$

where $\hat{x}_k \in R^n$ is the state vector of the filter; $\hat{z}_k \in R^q$ is the estimated signal of z_k. And

$$\begin{aligned} A_{f\sigma_k h} &= \sum_{i=1}^{r} h_{ai} A_{f\sigma_k i}, \; B_{f\sigma_k h} = \sum_{i=1}^{r} h_{ai} B_{f\sigma_k i}, \\ E_{f\sigma_k h} &= \sum_{i=1}^{r} h_{ai} E_{f\sigma_k i}. \end{aligned}$$

The filter matrices $A_{f\sigma_k i}$, $B_{f\sigma_k i}$, and $E_{f\sigma_k i}$ are to be determined. The variable σ_k is the discrete-time Markov chain, which takes a value in $\mathsf{V} = \{1, 2, \ldots, V\}$ with the following transition probability matrix $\Gamma^{\rho_{k+1}} = [\mu_{cd}^{\rho_{k+1}}]$ and

$$\Pr\{\sigma_{k+1} = d | \sigma_k = c\} = \mu_{cd}^{\rho_{k+1}}. \tag{9.8}$$

It is easily observed that for all $c, d \in \mathsf{V}$, $0 \leq \mu_{cd}^{\rho_{k+1}} \leq 1$ and $\sum_{d=1}^{V} \mu_{cd}^{\rho_{k+1}} = 1$ hold.

Remark 9.1 Note that MJSs play an important role in modeling practical systems whose parameters change by obeying certain transition probabilities. In this chapter, it is assumed that the transition probability is known, which may be obtained by conducting statistical analysis on enough samples from practical experiments. Some works have been reported on how to achieve the transition probability, including the recursive Kullback–Leibler estimation approach [6] and the online Bayesian estimation method [7].

Define $e_k = z_k - \hat{z}_k$ and $\xi_k = \left[x_k^T \; \hat{x}_k^T\right]^T$. The following filtering error system is obtained from (9.3), (9.5) and (9.7):

$$\begin{cases} \xi_{k+1} = \bar{A}_{ach}\xi_k + \bar{B}_{ach}w_k, \\ e_k = \bar{E}_{ach}\xi_k, \end{cases} \tag{9.9}$$

where

$$\bar{A}_{ach} = \sum_{i=1}^{r}\sum_{j=1}^{r} h_{ai}h_{aj}\bar{A}_{acij}, \; \bar{A}_{acij} = \begin{bmatrix} A_{ai} & 0 \\ B_{fcj}SC_{ai} & A_{fcj} \end{bmatrix},$$

$$\bar{B}_{ach} = \sum_{i=1}^{r}\sum_{j=1}^{r} h_{ai}h_{aj}\bar{B}_{acij}, \; \bar{B}_{acij} = \begin{bmatrix} B_{ai} \\ B_{fcj}SD_{ai} \end{bmatrix},$$

$$\bar{E}_{ach} = \sum_{i=1}^{r} h_{ai}\bar{E}_{aci}, \; \bar{E}_{aci} = \left[E_{ai} \; -E_{fci}\right].$$

We introduce the following definition, which is helpful to analyze the main results in this chapter.

Definition 9.1 ([8, 9]) A finite Markov chain $\rho_k \in \mathsf{L}$ is said to be homogeneous (nonhomogeneous) if for all $k \geq 0$, the transition probability satisfies $\Pr\{\rho_{k+1} = b|\rho_k = a\} = \pi_{ab}$ (or π_{abk}), where π_{ab} (or π_{abk}) denotes a probability function.

Remark 9.2 From Definition 9.1, it is easy to find that ρ_k is a homogenous Markov chain. Since $\mu_{cd}^{\rho_{k+1}}$ in (9.8) is time-varying but keeps constant for the same ρ_{k+1}, σ_k is nonhomogenous and it stands for the finite piecewise homogeneous Markov jump.

In this chapter, our aim is to design an asynchronous filter in the form of (9.7) such that for a prescribed scalar $\gamma > 0$, the filtering error system (9.9) meets two requirements as follows:
(1) The system (9.9) is stochastically stable with $w_k \equiv 0$;
(2) Under the zero-initial condition for any nonzero $w_k \in l_2[0, \infty)$, e_k satisfies

$$\sup_k \sqrt{E\{||e_k||_2^2\}} < \gamma \sqrt{\sum_{k=0}^{\infty} ||w_k||_2^2}. \tag{9.10}$$

9.3 Main Results

To keep the filtering error system (9.9) stochastically stable and satisfy a given L_2–L_∞ performance index, we firstly propose a sufficient condition with the premise that the filter gains of the system (9.7) are known. Then, the solution to filter gains is given.

Theorem 9.3 *Consider the filtering error system (9.9). For a prescribed $\gamma > 0$, if there exist matrices $P_{ac} > 0$ and W_c for any $i,\ j \in R$, $a \in L$ and $c \in V$ satisfying*

$$\Xi_{acii} < 0, \tag{9.11}$$

$$\Xi_{acij} + \Xi_{acji} < 0,\ i < j, \tag{9.12}$$

$$\Upsilon_{aci} < 0, \tag{9.13}$$

where

$$\Xi_{acij} = \begin{bmatrix} Q_{ac} - W_c - W_c^T & W_c\bar{A}_{acij} & W_c\bar{B}_{acij} \\ * & -P_{ac} & 0 \\ * & * & -I \end{bmatrix},$$

$$\Upsilon_{aci} = \begin{bmatrix} -\gamma^2 I & \bar{E}_{aci} \\ * & -P_{ac} \end{bmatrix},\ Q_{ac} = \sum_{b=1}^{L}\sum_{d=1}^{V}\pi_{ab}\mu_{cd}^{b}P_{bd},$$

system (9.9) is stochastically stable with a given L_2–L_∞ performance γ.

Proof According to the fuzzy principle, from (9.9), we have

$$\begin{aligned}\Xi_{ash} =& \sum_{i=1}^{r}\sum_{j=1}^{r} h_{ai}h_{aj}\Xi_{acij} = \sum_{i=1}^{r} h_{ai}^2\Xi_{acii} \\ &+ \sum_{i=1}^{r-1}\sum_{j=i+1}^{r} h_{ai}h_{aj}(\Xi_{acij} + \Xi_{acji}) < 0,\end{aligned} \tag{9.14}$$

and

$$\Upsilon_{ach} = \sum_{i=1}^{r} h_{ai}\Upsilon_{aci} < 0, \tag{9.15}$$

where

$$\Xi_{ach} = \begin{bmatrix} Q_{ac} - W_c - W_c^T & W_c\bar{A}_{ach} & W_c\bar{B}_{ach} \\ * & -P_{ac} & 0 \\ * & * & -I \end{bmatrix},$$

$$\Upsilon_{ach} = \begin{bmatrix} -\gamma^2 I & \bar{E}_{ach} \\ * & -P_{ac} \end{bmatrix}.$$

Due to $P_{bd} > 0$, the following inequality is obtained:

$$(Q_{ac} - W_c)Q_{ac}^{-1}(Q_{ac} - W_c^T) \geq 0. \tag{9.16}$$

Hence,

$$Q_{ac} - W_c - W_c^T \geq -W_c Q_{ac}^{-1} W_c^T. \tag{9.17}$$

According to (9.14), it follows that

$$\begin{bmatrix} -W_c Q_{ac}^{-1} W_c^T & W_c \bar{A}_{ach} & W_c \bar{B}_{ach} \\ * & -P_{ac} & 0 \\ * & * & -I \end{bmatrix} < 0. \tag{9.18}$$

Pre-multiply diag$\{W_c^{-1}, I, I\}$ and post-multiply diag$\{W_c^{-T}, I, I\}$ to (9.18). It follows that

$$\begin{bmatrix} -Q_{ac}^{-1} & \bar{A}_{ach} & \bar{B}_{ach} \\ * & -P_{ac} & 0 \\ * & * & -I \end{bmatrix} < 0. \tag{9.19}$$

By applying Schur Complement to (9.19), it is easy to find that

$$G_1 = \bar{A}_{ach}^T Q_{ac} \bar{A}_{ach} - P_{ac} < 0, \tag{9.20}$$

and

$$G_2 = \begin{bmatrix} \bar{A}_{ach}^T \\ \bar{B}_{ach}^T \end{bmatrix} Q_{ac} \begin{bmatrix} \bar{A}_{ach}^T \\ \bar{B}_{ach}^T \end{bmatrix}^T - \begin{bmatrix} P_{ac} & 0 \\ * & I \end{bmatrix} < 0. \tag{9.21}$$

Now, we construct the following Lyapunov function:

$$V_k = \xi_k^T P_{ac} \xi_k. \tag{9.22}$$

By computing the difference of V_k and from (9.9), we obtain:

$$\begin{aligned} \Delta V_k &= V_{k+1} - V_k = \xi_{k+1}^T P_{bd} \xi_{k+1} - \xi_k^T P_{ac} \xi_k \\ &= \begin{bmatrix} \xi_k \\ w_k \end{bmatrix}^T \begin{bmatrix} \bar{A}_{ach}^T \\ \bar{B}_{ach}^T \end{bmatrix} P_{bd} \begin{bmatrix} \bar{A}_{ach}^T \\ \bar{B}_{ach}^T \end{bmatrix}^T \begin{bmatrix} \xi_k \\ w_k \end{bmatrix} - \xi_k^T P_{ac} \xi_k. \end{aligned} \tag{9.23}$$

Since

$$\Pr\{\rho_{k+1} = b | \rho_k = a, \sigma_k = c\} = \pi_{ab} \tag{9.24}$$

$$\begin{aligned} &\Pr\{\rho_{k+1} = b, \sigma_{k+1} = d | \rho_k = a, \sigma_k = c\} \\ &= \Pr\{\sigma_{k+1} = d | \rho_k = a, \sigma_k = c, \rho_{k+1} = b\} \\ &\times \Pr\{\rho_{k+1} = b | \rho_k = a, \sigma_k = c\} = \pi_{ab} \mu_{cd}^b, \end{aligned} \tag{9.25}$$

we obtain the following expectation of ΔV_k:

$$E\{\Delta V_k\} = \begin{bmatrix} \xi_k \\ w_k \end{bmatrix}^T \begin{bmatrix} \bar{A}_{ach}^T \\ \bar{B}_{ach}^T \end{bmatrix} Q_{ac} \begin{bmatrix} \bar{A}_{ach}^T \\ \bar{B}_{ach}^T \end{bmatrix}^T \begin{bmatrix} \xi_k \\ w_k \end{bmatrix} - \xi_k^T P_{ac} \xi_k. \tag{9.26}$$

From (9.20), $E\{\Delta V_k\} = \xi_k^T G_1 \xi_k < 0$ holds when $w_k \equiv 0$. In this case, we clearly find that system (9.9) is stochastically stable.

Then, we can verify that system (9.9) meets the given L_2–L_∞ performance index under the zero initial state.

Consider the following index function:

$$J = E\{\Delta V_k\} - E\{w_k^T w_k\}. \tag{9.27}$$

Accordingly, we have

$$J = \begin{bmatrix} \xi_k \\ w_k \end{bmatrix}^T G_2 \begin{bmatrix} \xi_k \\ w_k \end{bmatrix}. \tag{9.28}$$

Owing to $G_2 < 0$, it follows that

$$E\{\Delta V_k\} - E\{w_k^T w_k\} < 0. \tag{9.29}$$

Adding both sides of (9.29) as index from 0 to $k-1$ under the initial zero state, we have

$$E\{V_k\} - \sum_{v=0}^{k-1} w_v^T w_v < 0. \tag{9.30}$$

Through applying Schur Complement to (9.15), the following inequality holds:

$$\bar{E}_{ach}^T \bar{E}_{ach} < \gamma^2 P_{ac}. \tag{9.31}$$

Consequently, we obtain that

$$\begin{aligned} E\{e_k^T e_k\} &= E\{\xi_k^T \bar{E}_{ach}^T \bar{E}_{ach} \xi_k\} < \gamma^2 E\{\xi_k^T P_{ac} \xi_k\} \\ &< \gamma^2 \sum_{v=0}^{k-1} w_v^T w_v < \gamma^2 \sum_{v=0}^{\infty} w_v^T w_v, \end{aligned} \tag{9.32}$$

which implies that it meets the L_2–L_∞ performance index in (9.10). The proof is completed.

Now, we are in a position to obtain the filter parameters of system (9.7) on the basis of Theorem 9.3, as shown below.

Theorem 9.4 *Consider the filter system (9.7) and the filtering error system (9.9). For a prescribed* $\gamma > 0$, *if there exist matrices*

$$\begin{bmatrix} P_{1ac} & P_{2ac} \\ * & P_{3ac} \end{bmatrix} > 0, \begin{bmatrix} W_{1c} & Y_c \\ W_{2c} & Y_c \end{bmatrix},$$

$\hat{A}_{fci}$, $\hat{B}_{fci}$, $\hat{E}_{fci}$, *and the diagonal matrix* $M_{ac} > 0$ *for any* $i,\ j \in \mathcal{R}$, $a \in \mathcal{L}$ *and* $c \in \mathcal{V}$ *satisfying*

$$\Psi^1_{acii} = \begin{bmatrix} \Sigma^1_{acii} & \Sigma^2_{ci} & \Sigma^3_{ai} M_{ac} \\ * & -M_{ac} & 0 \\ * & * & -M_{ac} \end{bmatrix} < 0, \tag{9.33}$$

$$\Psi^2_{acij} = \begin{bmatrix} \Sigma^1_{acij} + \Sigma^1_{acji} & \Psi^2_{12cij} & \Psi^2_{13aij} M_{ac} \\ * & -\tilde{M}_{ac} & 0 \\ * & * & -\tilde{M}_{ac} \end{bmatrix} < 0,\ i < j, \tag{9.34}$$

$$\begin{bmatrix} -\gamma^2 I & E_{ai} & -\hat{E}_{fci} \\ * & -P_{1ac} & -P_{2ac} \\ * & * & -P_{3ac} \end{bmatrix} < 0, \tag{9.35}$$

where

$$\Sigma^1_{acij} = \begin{bmatrix} \Sigma^1_{11ac} & \Sigma^1_{12acij} & \Sigma^1_{13acij} \\ * & -\Sigma^1_{22ac} & 0 \\ * & * & -I \end{bmatrix},$$

$$\Sigma^1_{22ac} = \begin{bmatrix} P_{1ac} & P_{2ac} \\ * & P_{3ac} \end{bmatrix},\ \tilde{M}_{ac} = \{M_{ac}, M_{ac}\},$$

$$\Sigma^1_{11ac} = \begin{bmatrix} Q_{1ac} - W_{1c} - W^T_{1c} & Q_{2ac} - Y_c - W^T_{2c} \\ * & Q_{3ac} - Y_c - Y^T_c \end{bmatrix},$$

$$\Sigma^1_{12acij} = \begin{bmatrix} W_{1c} A_{ai} + \hat{B}_{fcj} \acute{S} C_{ai} & \hat{A}_{fcj} \\ W_{2c} A_{ai} + \hat{B}_{fcj} \acute{S} C_{ai} & \hat{A}_{fcj} \end{bmatrix},$$

$$\Sigma^1_{13acij} = \begin{bmatrix} W_{1c} B_{ai} + \hat{B}_{fcj} \acute{S} D_{ai} \\ W_{2c} B_{ai} + \hat{B}_{fcj} \acute{S} D_{ai} \end{bmatrix},$$

$$\Sigma^2_{cj} = \begin{bmatrix} \hat{B}^T_{fcj} & \hat{B}^T_{fcj} & 0 & 0 & 0 \end{bmatrix}^T,\ \Sigma^3_{ai} = \begin{bmatrix} 0 & 0 & \grave{S} C_{ai} & 0 & \grave{S} D_{ai} \end{bmatrix}^T,$$

$$\Psi^2_{12cij} = \begin{bmatrix} \Sigma^2_{cj} & \Sigma^2_{ci} \end{bmatrix},\ \Psi^2_{13aij} = \begin{bmatrix} \Sigma^3_{ai} & \Sigma^3_{aj} \end{bmatrix},$$

$$\begin{bmatrix} Q_{1ac} & Q_{2ac} \\ * & Q_{3ac} \end{bmatrix} = \sum_{b=1}^{L} \sum_{d=1}^{V} \pi_{ab} \mu^b_{cd} \begin{bmatrix} P_{1bd} & P_{2bd} \\ * & P_{3bd} \end{bmatrix},$$

system (9.9) is stochastically stable with a given L_2–L_∞ performance γ. Furthermore, the parameters of the filter in (9.7) can be achieved by

$$\left[\begin{array}{c|c} A_{fci} & B_{fci} \\ \hline E_{fci} & - \end{array}\right] = \left[\begin{array}{c|c} Y_c^{-1}\hat{A}_{fci} & Y_c^{-1}\hat{B}_{fci} \\ \hline \hat{E}_{fci} & - \end{array}\right]. \tag{9.36}$$

Proof Through Schur Complement, it follows from (9.33) that

$$\Sigma_{acii}^1 + \Sigma_{ci}^2 M_{ac}^{-1}(\Sigma_{ci}^2)^T + \Sigma_{ai}^3 M_{ac}(\Sigma_{ai}^3)^T < 0. \tag{9.37}$$

From (9.6), we have $\Lambda^T\Lambda < \grave{S}^T\grave{S}$. Therefore, we obtain

$$\Sigma_{acii}^1 + \Sigma_{ci}^2 M_{ac}^{-1}(\Sigma_{ci}^2)^T + \Sigma_{ai}^{3'} M_{ac}(\Sigma_{ai}^{3'})^T < 0, \tag{9.38}$$

where

$$\Sigma_{ai}^{3'} = \left[0\ 0\ \Lambda C_{ai}\ 0\ \Lambda D_{ai}\right]^T.$$

Due to the elementary inequality $x^Ty + y^Tx \le \varepsilon x^Tx + \varepsilon^{-1}y^Ty$, the following inequality is acquired:

$$\Psi_{acii}^{1'} = \Sigma_{acii}^1 + \Sigma_{ci}^2(\Sigma_{ai}^{3'})^T + \Sigma_{ai}^{3'}(\Sigma_{ci}^2)^T < 0, \tag{9.39}$$

where

$$\Psi_{acij}^{1'} = \begin{bmatrix} \Sigma_{11ac}^1 & \Sigma_{12acij}^{1'} & \Sigma_{13acij}^{1'} \\ * & -\Sigma_{22ac}^1 & 0 \\ * & * & -I \end{bmatrix},$$
$$\Sigma_{12acij}^{1'} = \begin{bmatrix} W_{1c}A_{ai} + \hat{B}_{fcj}SC_{ai} & \hat{A}_{fci} \\ W_{2c}A_{ai} + \hat{B}_{fcj}SC_{ai} & \hat{A}_{fci} \end{bmatrix},$$
$$\Sigma_{13acij}^{1'} = \begin{bmatrix} W_{1c}B_{ai} + \hat{B}_{fcj}SD_{ai} \\ W_{2c}B_{ai} + \hat{B}_{fcj}SD_{ai} \end{bmatrix},\ i = j.$$

Let

$$P_{ac} = \begin{bmatrix} P_{1ac} & P_{2ac} \\ * & P_{3ac} \end{bmatrix},\ W_c = \begin{bmatrix} W_{1c} & Y_c \\ W_{2c} & Y_c \end{bmatrix},$$
$$\hat{A}_{fci} = Y_cA_{fci},\ \hat{B}_{fci} = Y_cB_{fci},\ \hat{E}_{fci} = E_{fci} \tag{9.40}$$

and from (9.9), we can find that (9.39) and (9.35) are equivalent to (9.11) and (9.13), respectively. Furthermore, it is evident that we can obtain (9.11) from (9.33), and via the similar proof process, we can derive (9.12) from (9.34) as well. If there is a solution to (9.33)–(9.35), the filter parameters will be achieved by solving (9.36). Hence, the proof is completed.

Remark 9.5 In designing filter parameters, the slack matrix $W_c = \begin{bmatrix} W_{1c} & Y_c \\ W_{2c} & Y_c \end{bmatrix}$ is introduced to eliminate nonlinear couplings in Theorem 9.3. This is a conventional approach, which has been used by many published works such as the nonfragile distributed filter design [10] and the event-triggered H_∞ filter [11]. It is worth noting that some unnecessary matrices W_{1c} and W_{2c} increase the number of unknown variables and bring more conservatism. Motivated by [12], another approach with less conservatism is developed to solve filter gains by removing W_{1c} and W_{2c}, shown in Theorem 9.6.

Theorem 9.6 *Consider the filter system (9.7) and the filtering error system (9.9). For a prescribed $\gamma > 0$, if there exist matrices*

$$\begin{bmatrix} P_{1ac} & P_{2ac} \\ * & P_{3ac} \end{bmatrix} > 0,$$

Y_c, $\hat{A}_{fci}$, $\hat{B}_{fci}$, $\hat{E}_{fci}$, *and the diagonal matrix* $M_{ac} > 0$ *for any* $i,\ j \in \mathsf{R}$, $a \in \mathsf{L}$ *and* $c \in \mathsf{V}$ *satisfying (9.35) and*

$$N_1^T \Omega_{acii}^1 N_1 < 0, \tag{9.41}$$

$$H_1^T \Omega_{acii}^1 H_1 < 0, \tag{9.42}$$

$$N_2^T \Omega_{acij}^2 N_2 < 0,\ i < j, \tag{9.43}$$

$$H_2^T \Omega_{acij}^2 H_2 < 0,\ i < j, \tag{9.44}$$

where

$$\Omega_{acii}^1 = \begin{bmatrix} \Phi_{acii}^1 & \Sigma_{ci}^2 & \Sigma_{ai}^3 M_{ac} \\ * & -M_{ac} & 0 \\ * & * & -M_{ac} \end{bmatrix},$$

$$\Omega_{acij}^2 = \begin{bmatrix} \Phi_{acij}^1 + \Phi_{acji}^1 & \Psi_{12cij}^2 & \Psi_{13aij}^2 M_{ac} \\ * & -\tilde{M}_{ac} & 0 \\ * & * & -\tilde{M}_{ac} \end{bmatrix},$$

$$\Phi_{acij}^1 = \begin{bmatrix} \Phi_{11ac}^1 & \Phi_{12acij}^2 & \Phi_{13acij}^1 \\ * & -\Sigma_{22ac}^1 & 0 \\ * & * & -I \end{bmatrix},$$

$$\Phi_{12acij}^2 = \begin{bmatrix} \hat{B}_{fcj} \acute{S} C_{ai} & \hat{A}_{fcj} \\ \hat{B}_{fcj} \acute{S} C_{ai} & \hat{A}_{fcj} \end{bmatrix},\quad \Phi_{13acij}^1 = \begin{bmatrix} \hat{B}_{fcj} \acute{S} D_{ai} \\ \hat{B}_{fcj} \acute{S} D_{ai} \end{bmatrix},$$

$$\Phi_{11ac}^1 = \begin{bmatrix} Q_{1ac} & Q_{2ac} - Y_c \\ * & Q_{3ac} - Y_c - Y_c^T \end{bmatrix},$$

$$N_1 = \begin{bmatrix} N_1^1 \\ N_1^2 \end{bmatrix}, \; N_2 = \begin{bmatrix} N_2^1 \\ N_2^2 \end{bmatrix}, \; H_1 = \begin{bmatrix} 0 \\ 0 \\ H_1^3 \end{bmatrix}, \; H_2 = \begin{bmatrix} 0 \\ 0 \\ H_2^3 \end{bmatrix},$$
$$N_1^1 = \begin{bmatrix} 0 \; A_{ai} \; 0 \; B_{ai} \; 0 \; 0 \end{bmatrix}, \; N_1^2 = diag\{I, I, I, I, I, I\},$$
$$N_2^1 = \begin{bmatrix} 0 \; \frac{A_{ai}+A_{aj}}{2} \; 0 \; \frac{B_{ai}+B_{aj}}{2} \; 0 \; 0 \; 0 \; 0 \end{bmatrix},$$
$$N_2^2 = diag\{I, I, I, I, I, I, I, I\},$$
$$H_1^3 = diag\{I, I, I, I, I\}, \; H_2^3 = diag\{I, I, I, I, I, I, I\},$$

system (9.9) is stochastically stable with a given L_2–L_∞ performance γ. Furthermore, the parameters of the filter in (9.7) can be achieved via (9.36).

Proof From (9.33), we obtain that

$$\Psi_{acii}^1 = \Omega_{acii}^1 + \tilde{H}_1^T \begin{bmatrix} W_{1c} \\ W_{2c} \end{bmatrix} \tilde{N}_1 + \tilde{N}_1^T \begin{bmatrix} W_{1c} \\ W_{2c} \end{bmatrix}^T \tilde{H}_1 < 0, \tag{9.45}$$

where

$$\tilde{N}_1 = \begin{bmatrix} -I \; 0 \; A_{ai} \; 0 \; B_{ai} \; 0 \; 0 \end{bmatrix},$$
$$\tilde{H}_1 = \begin{bmatrix} I \; 0 \; 0 \; 0 \; 0 \; 0 \; 0 \\ 0 \; I \; 0 \; 0 \; 0 \; 0 \; 0 \end{bmatrix}.$$

By calculating, we have that the orthogonal complements of $\tilde{N}_1^T$ and $\tilde{H}_1^T$ are N_1 and H_1, respectively. From Finsler's lemma, it follows that when (9.41) and (9.42) hold, (9.33) will hold. By the similar method, we can find that the feasibility of (9.43) and (9.44) is equivalent to that of (9.34). Therefore, the proof is completed.

Remark 9.7 For a given $\gamma > 0$, the number of decision variables is $(n + m + q)nrV + (2L + 3)n^2V + (n + m)LV$ in Theorem 9.4. With using Finsler's lemma, W_{1c} and W_{2c} are eliminated and the number of decision variables decreases to $(n + m + q)nrV + (2L + 1)n^2V + (n + m)LV$ in Theorem 9.6. However, there are $(2 + r)rLV$ LMIs in Theorem 9.6, which is $0.5(1 + r)rLV$ LMIs more than that in Theorem 9.4. Hence, compared with Theorem 9.4, Theorem 9.6 brings less conservatism but has more computation burden.

9.4 Illustrative Example

In this section, one example is provided to demonstrate the correctness and effectiveness of the proposed methods.

Consider the system (9.3) with two fuzzy rules and three Markov jump modes. And the parameters are given as follows:

$$\left[\begin{array}{c|c} A_{11} & B_{11} \\ \hline C_{11} & D_{11} \\ \hline E_{11} & - \end{array}\right] = \left[\begin{array}{cc|c} 0.58 & 0.39 & 0.11 \\ -0.12 & 0.45 & 0.19 \\ \hline 1 & 0 & 1.1 \\ \hline 1 & 0 & - \end{array}\right],$$

$$\left[\begin{array}{c|c} A_{12} & B_{12} \\ \hline C_{12} & D_{12} \\ \hline E_{12} & - \end{array}\right] = \left[\begin{array}{cc|c} 0.82 & 0.12 & 0.21 \\ -0.20 & 0.60 & 0.29 \\ \hline 1 & 0 & 0.9 \\ \hline 1 & 0 & - \end{array}\right],$$

$$\left[\begin{array}{c|c} A_{21} & B_{21} \\ \hline C_{21} & D_{21} \\ \hline E_{21} & - \end{array}\right] = \left[\begin{array}{cc|c} 0.65 & 0.48 & 0.30 \\ -0.16 & 0.52 & 0.20 \\ \hline 1 & 0 & 1.05 \\ \hline 1 & 0 & - \end{array}\right],$$

$$\left[\begin{array}{c|c} A_{22} & B_{22} \\ \hline C_{22} & D_{22} \\ \hline E_{22} & - \end{array}\right] = \left[\begin{array}{cc|c} 0.88 & 0.13 & 0.10 \\ -0.13 & 0.70 & 0.20 \\ \hline 1 & 0 & 0.95 \\ \hline 1 & 0 & - \end{array}\right],$$

$$\left[\begin{array}{c|c} A_{31} & B_{31} \\ \hline C_{31} & D_{31} \\ \hline E_{31} & - \end{array}\right] = \left[\begin{array}{cc|c} 0.65 & 0.39 & 0.15 \\ -0.20 & 0.42 & 0.24 \\ \hline 1 & 0 & 1.07 \\ \hline 1 & 0 & - \end{array}\right],$$

$$\left[\begin{array}{c|c} A_{32} & B_{32} \\ \hline C_{32} & D_{32} \\ \hline E_{32} & - \end{array}\right] = \left[\begin{array}{cc|c} 0.85 & 0.10 & 0.25 \\ -0.15 & 0.65 & 0.30 \\ \hline 1 & 0 & 0.98 \\ \hline 1 & 0 & - \end{array}\right].$$

We assume that normalized fuzzy weighting functions are:

$$\begin{cases} h_{11} = 0.5(sin(w_k))^2, \ h_{a2} = 1 - h_{a1}, \\ h_{21} = 0.6(cos(w_k))^2, \ a \in \{1, 2, 3\}, \\ h_{31} = 0.2(cos(w_k))^{0.4}. \end{cases}$$

The transition probability matrix of the studied plant is given as

$$\Pi = \begin{bmatrix} 0.3 & 0.4 & 0.3 \\ 0.4 & 0.2 & 0.4 \\ 0.25 & 0.45 & 0.3 \end{bmatrix}.$$

The designed asynchronous filter is supposed to have three operation modes, obeying the following transition probability matrices:

$$\Gamma^1 = \begin{bmatrix} 0.25 & 0.45 & 0.3 \\ 0.3 & 0.3 & 0.4 \\ 0.25 & 0.45 & 0.3 \end{bmatrix}, \quad \Gamma^2 = \begin{bmatrix} 0.35 & 0.2 & 0.45 \\ 0.25 & 0.35 & 0.4 \\ 0.25 & 0.45 & 0.3 \end{bmatrix},$$

$$\Gamma^3 = \begin{bmatrix} 0.35 & 0.25 & 0.4 \\ 0.35 & 0.35 & 0.3 \\ 0.45 & 0.25 & 0.3 \end{bmatrix}.$$

The sensor failure is chosen as $0.85 < S < 0.95$, which means that $\acute{S} = 0.9$ and $\grave{S} = 0.05$. Theorems 9.4 and 9.6 are utilized to obtain the filter matrices via Matlab respectively, as follows:
Theorem 9.4:

$$\left[\begin{array}{c|c} A_{f11} & B_{f11} \\ \hline E_{f11} & - \end{array}\right] = \left[\begin{array}{cc|c} 0.4274 & 0.4145 & -0.2171 \\ -0.3579 & 0.4581 & -0.2176 \\ \hline -0.9884 & -0.0034 & - \end{array}\right],$$

$$\left[\begin{array}{c|c} A_{f12} & B_{f12} \\ \hline E_{f12} & - \end{array}\right] = \left[\begin{array}{cc|c} 0.6858 & 0.1505 & -0.2004 \\ -0.4364 & 0.6504 & -0.3152 \\ \hline -0.9884 & -0.0034 & - \end{array}\right],$$

$$\left[\begin{array}{c|c} A_{f21} & B_{f21} \\ \hline E_{f21} & - \end{array}\right] = \left[\begin{array}{cc|c} 0.4201 & 0.4068 & -0.2176 \\ -0.3600 & 0.4588 & -0.2176 \\ \hline -0.9881 & -0.0051 & - \end{array}\right],$$

$$\left[\begin{array}{c|c} A_{f22} & B_{f22} \\ \hline E_{f22} & - \end{array}\right] = \left[\begin{array}{cc|c} 0.6841 & 0.1497 & -0.2003 \\ -0.4373 & 0.6468 & -0.3121 \\ \hline -0.9881 & -0.0051 & - \end{array}\right],$$

$$\left[\begin{array}{c|c} A_{f31} & B_{f31} \\ \hline E_{f31} & - \end{array}\right] = \left[\begin{array}{cc|c} 0.4281 & 0.4176 & -0.2170 \\ -0.3568 & 0.4558 & -0.2180 \\ \hline -0.9864 & -0.0017 & - \end{array}\right],$$

$$\left[\begin{array}{c|c} A_{f32} & B_{f32} \\ \hline E_{f32} & - \end{array}\right] = \left[\begin{array}{cc|c} 0.6811 & 0.1487 & -0.2016 \\ -0.4321 & 0.6551 & -0.3104 \\ \hline -0.9864 & -0.0017 & - \end{array}\right],$$

and the optimal $\gamma^* = 0.1597$.
Theorem 9.6:

$$\left[\begin{array}{c|c} A_{f11} & B_{f11} \\ \hline E_{f11} & - \end{array}\right] = \left[\begin{array}{cc|c} 0.4255 & 0.4113 & -0.2177 \\ -0.3597 & 0.4598 & -0.2179 \\ \hline -0.9892 & -0.0028 & - \end{array}\right],$$

$$\left[\begin{array}{c|c} A_{f12} & B_{f12} \\ \hline E_{f12} & - \end{array}\right] = \left[\begin{array}{cc|c} 0.6857 & 0.1517 & -0.2006 \\ -0.4397 & 0.6491 & -0.3178 \\ \hline -0.9892 & -0.0028 & - \end{array}\right],$$

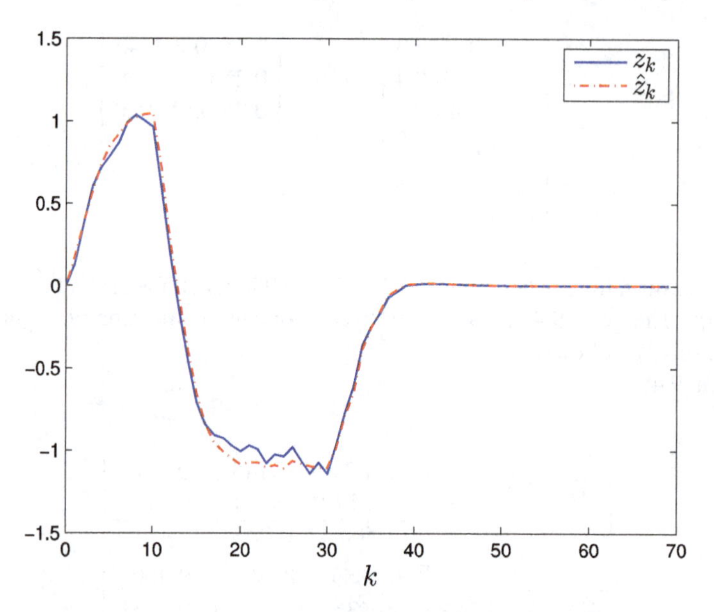

Fig. 9.2 Original output z_k and estimated output $\hat{z}_k$ by Theorem 9.4

$$\left[\begin{array}{c|c} A_{f21} & B_{f21} \\ \hline E_{f21} & - \end{array}\right] = \left[\begin{array}{cc|c} 0.4189 & 0.4030 & -0.2181 \\ -0.3619 & 0.4600 & -0.2181 \\ \hline -0.9890 & -0.0032 & - \end{array}\right],$$

$$\left[\begin{array}{c|c} A_{f22} & B_{f22} \\ \hline E_{f22} & - \end{array}\right] = \left[\begin{array}{cc|c} 0.6842 & 0.1503 & -0.2002 \\ -0.4408 & 0.6448 & -0.3148 \\ \hline -0.9890 & -0.0032 & - \end{array}\right],$$

$$\left[\begin{array}{c|c} A_{f31} & B_{f31} \\ \hline E_{f31} & - \end{array}\right] = \left[\begin{array}{cc|c} 0.4264 & 0.4129 & -0.2176 \\ -0.3582 & 0.4584 & -0.2180 \\ \hline -0.9877 & 0.0011 & - \end{array}\right],$$

$$\left[\begin{array}{c|c} A_{f32} & B_{f32} \\ \hline E_{f32} & - \end{array}\right] = \left[\begin{array}{cc|c} 0.6815 & 0.1480 & -0.2015 \\ -0.4355 & 0.6541 & -0.3132 \\ \hline -0.9877 & 0.0011 & - \end{array}\right],$$

and the optimal $\gamma^* = 0.1593$. Figures 9.2 and 9.3 present the responses of z_k and $\hat{z}_k$ and the filtering error e_k by Theorem 9.4, respectively. And Figs. 9.4 and 9.5 show the corresponding results via Theorem 9.6. It is easy to find that e_k approximates zero as time k passes by and we can conclude that our proposed method is correct and effective.

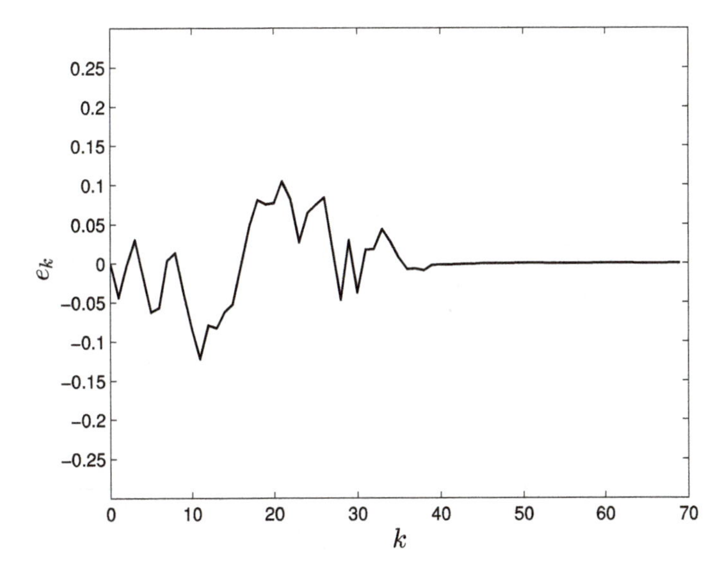

Fig. 9.3 Filtering error e_k by Theorem 9.4

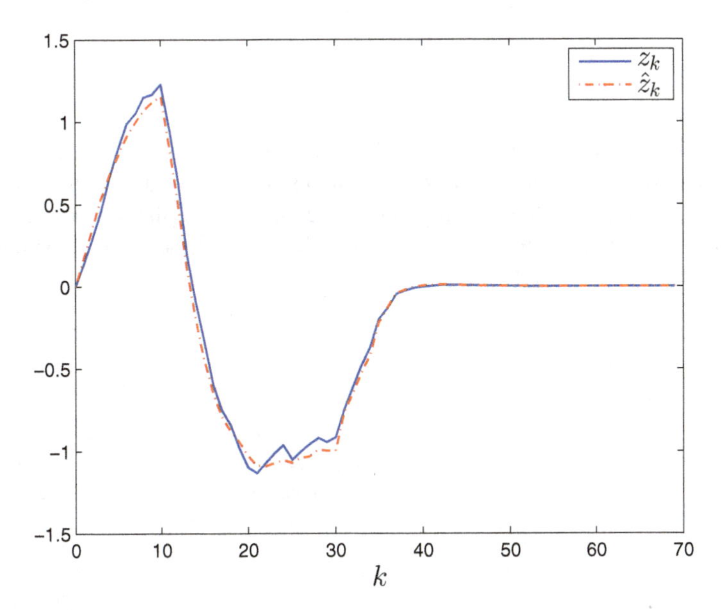

Fig. 9.4 Original output z_k and estimated output $\hat{z}_k$ by Theorem 9.6

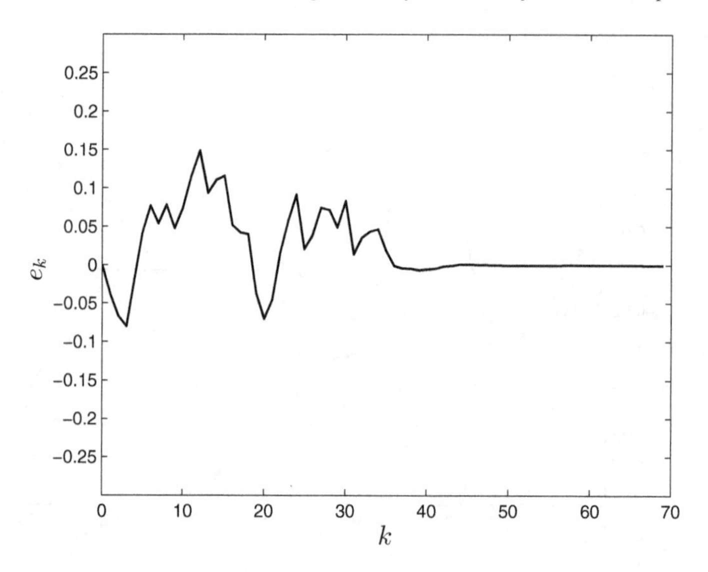

Fig. 9.5 Filtering error e_k by Theorem 9.6

9.5 Conclusion

The reliable and asynchronous L_2–L_∞ filtering design issue has been addressed for fuzzy MJSs subject to sensor failures and mode asynchronization. By applying the Lyapunov function and the relaxing matrix technique, two filter design approaches have been obtained in the form of LMIs and filter matrices can be achieved by applying LMI Toolbox in Matlab.

References

1. Tian, E., Yue, D.: Reliable H_∞ filter design for T-S fuzzy model-based networked control systems with random sensor failure. Int. J. Robust Nonliner Control **23**(1), 15–32 (2013)
2. Wei, G., Han, F., Wang, L., Song, Y.: Reliable H_∞ filtering for discrete piecewise linear systems with inifinite distributed delays. Int. J. Gen. Syst. **43**(3), 346–358 (2014)
3. Sun, C., Wang, F., He, X.: Robust fault-tolerant control for fuzzy delay systems with unmeasurable premise variables via uncertain system approach. Int. J. Innov. Comput. Inf. Control **13**(3), 823–846 (2017)
4. Su, X., Shi, P., Wu, L., Basin, M.V.: Reliable filtering with strict dissipativity for T-S fuzzy time-delay systems. IEEE Trans. Cybern. **44**(12), 2470–2483 (2014)
5. Yang, G.-H., Dan, Y.: Adaptive reliable H_∞ filtering against sensor failures. IEEE Trans. Signal Process. **55**(7), 3161–3171 (2007)
6. Orguner, U., Demirekler, M.: An online sequential algorithm for the estimation of transition probabilities for jump Markov linear systems. Automatica **42**(10), 1735–1744 (2006)
7. Jilkov, V.P., Li, X.R.: Online Bayesian estimation of transition probabilities for Markovian jump systems. IEEE Trans. Signal Process. **52**(6), 1620–1630 (2004)

8. Marius, I.: Finite Markov Processes and Their Applications. Courier Corporation (1980)
9. Zhang, L.: H_∞ estimation for discrete-time piecewise homogeneous Markov jump linear systems. Automatica **45**(11), 2570–2576 (2009)
10. Zhang, D., Cai, W., Xie, L., Wang, Q.-G.: Nonfragile distributed filtering for T-S fuzzy systems in sensor networks. IEEE Trans. Fuzzy Syst. **23**(5), 1883–1890 (2015)
11. Zhang, C., Hu, J., Qiu, J., Chen, Q.: Event-triggered nonsynchronized H_∞ filtering for discrete-time T-S fuzzy systems based on piecewise Lyapunov functions. IEEE Trans. Syst., Man, Cybern.: Syst. **47**(8), 2330–2341 (2017)
12. Li, X., Lam, J., Gao, H., Li, P.: Improved results on H_∞ model reduction for Markovian jump systems with partly known transition probabilities. Syst. Control Lett. **70**, 109–117 (2014)

Chapter 10
Reliable Filtering of Nonlinear Markovian Jump Systems: The Continuous-Time Case

10.1 Introduction

In this chapter, the reliable L_2–L_∞ asynchronous filtering design issue is studied for the nonlinear continuous-time MJSs, which is modeled by the T–S fuzzy principle. Via the HMM theory, the asynchronous sensor and filter are designed: one is applied to represent the stochastic failure and the other is to estimate the plant. Although they are asynchronous with the plant via the same approach, both are conditionally independent. Based on the stochastic Lypunov function approach and the LMI technique, a sufficient condition is developed to guarantee the existence of the designed filter and satisfy the given L_2–L_∞ disturbance attenuation level of the filtering error systems in terms of LMIs.

10.2 Preliminary Analysis

Consider the following continuous-time T–S fuzzy systems with Markov jump:
Plant rule i: IF $\vartheta_1(t)$ is η_{i1}, $\vartheta_2(t)$ is η_{i2}, ..., and $\vartheta_p(t)$ is η_{ip}, THEN

$$\begin{cases} \dot{x}(t) = A_{r(t)i}x(t) + B_{r(t)i}w(t), \\ y(t) = C_{r(t)i}x(t) + D_{r(t)i}w(t), \\ z(t) = E_{r(t)i}x(t) + F_{r(t)i}w(t), \end{cases} \tag{10.1}$$

where η_{ij} ($i \in \mathcal{I} = \{1, 2, \ldots, r\}$, $j \in \{1, 2, \ldots, p\}$) is the fuzzy set with r fuzzy rules, and $\vartheta_j(t)$ is the premise variable. These variables $x(t) \in R^{n_x}$, $y(t) \in R^{n_y}$, $z(t) \in R^{n_z}$ and $w(t) \in R^{n_w}$ are the state, the measured output, the signal to be estimated and the disturbance belonging to $l_2[0, +\infty)$, respectively. Matrices $A_{r(t)i}$, $B_{r(t)i}$, $C_{r(t)i}$, $D_{r(t)i}$, $E_{r(t)i}$, and $F_{r(t)i}$ are given with appropriate dimensions. The variable $r(t)$ stands for the time-homogeneous Morkov jump process with right

S. Dong et al., *Control and Filtering of Fuzzy Systems with Switched Parameters*, Studies in Systems, Decision and Control 268,
https://doi.org/10.1007/978-3-030-35566-1_10

continuous trajectories. It takes values in $\mathcal{Q} = \{1, 2, \ldots, Q\}$ and is subject to the transition rate matrix $\Pi = [\pi_{de}]$ with

$$\Pr\{r(t+\Delta t) = e | r(t) = d\} = \begin{cases} \pi_{de}\Delta t + o(\Delta t), & d \neq e, \\ 1 + \pi_{ee}\Delta t + o(\Delta t), & d = e, \end{cases} \tag{10.2}$$

where π_{de} represents the jump rate from mode d at time t to mode e after Δt with $\pi_{de} \geq 0$ $(d \neq e)$, $\pi_{ee} = -\sum_{e=1, e\neq d}^{Q} \pi_{de}$ and $\sum_{e=1}^{Q} \pi_{de} = 0$. It is worth pointing out that Δt is the infinitesimal transition time interval, which means that $\lim_{\Delta t \to 0} \frac{o(\Delta t)}{\Delta t} = 0$.

Through the T–S fuzzy approach with $r(t) = d$, the overall fuzzy system is inferred as

$$\begin{cases} \dot{x}(t) = A_{dh}x(t) + B_{dh}w(t), \\ y(t) = C_{dh}x(t) + D_{dh}w(t), \\ z(t) = E_{dh}x(t) + F_{dh}w(t), \end{cases} \tag{10.3}$$

where

$$A_{dh} = \sum_{i=1}^{r} h_i(\vartheta(t))A_{di}, \ B_{dh} = \sum_{i=1}^{r} h_i(\vartheta(t))B_{di},$$

$$C_{dh} = \sum_{i=1}^{r} h_i(\vartheta(t))C_{di}, \ D_{dh} = \sum_{i=1}^{r} h_i(\vartheta(t))C_{di},$$

$$E_{dh} = \sum_{i=1}^{r} h_i(\vartheta(t))E_{di}, \ F_{dh} = \sum_{i=1}^{r} h_i(\vartheta(t))F_{di},$$

$$h_i(\vartheta(t)) = \frac{\prod_{j=1}^{p} \eta_{ij}(\vartheta_j(t))}{\sum_{i=1}^{r} \prod_{j=1}^{p} \eta_{ij}(\vartheta_j(t))}, \ \vartheta(t) = [\vartheta_1(t), \vartheta_1(t), \ldots, \vartheta_p(t)].$$

Here, $h_i(\vartheta(t))$ is the normalized fuzzy weighting function and $\eta_{ij}(\vartheta_j(t))$ is the grade of membership $\vartheta_j(t)$ in η_{ij}. It is supposed that $\prod_{j=1}^{p} \eta_{ij}(\vartheta_j(t)) \geq 0$ for all t. Accordingly, we clearly find that $h_i(\vartheta(t)) \geq 0$ and $\sum_{i=1}^{r} h_i(\vartheta(t)) = 1$.

In the following, $h_i(\vartheta(t))$ is written as h_i for brevity.

Considering that sensor failures occur randomly in control systems, we adopt the following stochastic failure model:

$$\hat{y}_i(t) = s_{\varphi(t)i}(t)y_i(t), \ i \in \{1, 2, \ldots, n_y\}, \tag{10.4}$$

where $\hat{y}_i(t)$ is the accepted signal from the ith sensor. The variable $\varphi(t)$ $(\varphi(t) \in \mathcal{L} = \{1, 2, \ldots, L\})$ is employed to describe the stochastic failure phenomenon, which applies the HMM principle with the conditional probability matrix $\Psi = [\varphi_{dl}]$ $(\varphi_{dl} \geq 0)$ and

$$\Pr\{\varphi(t) = l | r(t) = d\} = \varphi_{dl}, \tag{10.5}$$

where $\sum_{l=1}^{L} \varphi_{dl} = 1$.

In addition, the variable $s_{li}(t)$ ($\varphi(t) = l$) describes the ith failure level under the lth mode, satisfying

$$0 \leq \underline{s}_{li} \leq s_{li}(t) \leq \overline{s}_{li} \leq 1, \tag{10.6}$$

where $\underline{s}_{li}$ and $\overline{s}_{li}$ are given lower and upper bounds of s_{li}, respectively. Thus, $s_{li}(t)$ can be inferred as

$$s_{li}(t) = \hat{s}_{li} + \delta_{li}(t), \tag{10.7}$$

where

$$\hat{s}_{li} = \frac{\underline{s}_{li} + \overline{s}_{li}}{2}.$$

Since

$$\begin{aligned} -\check{s}_{li} \leq \delta_{li}(t) &= \frac{1}{2}(2s_{li}(t) - \overline{s}_{li} - \underline{s}_{li}) \leq \check{s}_{li}, \\ \check{s}_{li} &= \frac{\overline{s}_{li} - \underline{s}_{li}}{2}, \end{aligned} \tag{10.8}$$

we can conclude that $|\delta_{li}(t)| \leq \check{s}_{li} \leq 1$ and

$$\hat{y}(t) = S_l(t)y(t) = (\hat{S}_l + \Delta_l(t))y(t), \tag{10.9}$$

where

$$\begin{aligned} &\Delta_l(t) = \text{diag}\{\delta_{l1}(t), \delta_{l2}(t), \ldots, \delta_{ln_y}(t)\}, \ \Delta_l(t)\Delta_l(t) \leq \check{S}_l\check{S}_l \leq I, \\ &\hat{S}_l = \text{diag}\{\hat{s}_{l1}, \hat{s}_{l2}, \ldots, \hat{s}_{ln_y}\}, \ \check{S}_l = \text{diag}\{\check{s}_{l1}, \check{s}_{l2}, \ldots, \check{s}_{ln_y}\}. \end{aligned}$$

In this chapter, we are also interested in applying the HMM principle to design an asynchronous filter for tracking the plant. $\nu(t)$ ($\nu(t) \in \mathcal{G} = \{1, 2, \ldots, G\}$) is used to denote this situation and has identical features with $\varphi(t)$, satisfying the conditional probability matrix $\Phi = [\phi_{dg}]$ ($\phi_{dg} \geq 0$) and

$$\Pr\{\nu(t) = g | r(t) = d\} = \phi_{dg}, \tag{10.10}$$

where $\sum_{g=1}^{G} \phi_{dg} = 1$.

The following asynchronously reliable filter is devised to track the system (10.3) with $\nu(t) = g$:

$$\begin{cases} \dot{\hat{x}}(t) = \hat{A}_{gh}\hat{x}(t) + \hat{B}_{gh}\hat{y}(t), \\ \hat{z}(t) = \hat{E}_{gh}\hat{x}(t), \end{cases} \tag{10.11}$$

where

$$\hat{A}_{gh} = \sum_{i=1}^{r} h_i \hat{A}_{gi}, \ \hat{B}_{gh} = \sum_{i=1}^{r} h_i \hat{B}_{gi}, \ \hat{E}_{gh} = \sum_{i=1}^{r} h_i \hat{E}_{gi}.$$

Defining $e(t) = z(t) - \hat{z}(t)$ and $\xi(t) = \left[x^T(t) \ \hat{x}^T(t)\right]^T$, considering (10.3), (10.9) and (10.11), we have the filtering error system:

$$\begin{cases} \dot{\xi}(t) = (\bar{A}_{dlgh} + \bar{B}_{1gh}\Delta_l(t)\bar{C}_{dh})\xi(t) + (\bar{B}_{2dlgh} + \bar{B}_{1gh}\Delta_l(t)\bar{D}_{dh})w(t), \\ e(t) = \bar{E}_{dgh}\xi(t) + \bar{F}_{dh}w(t), \end{cases} \tag{10.12}$$

where

$$\bar{A}_{dlgh} = \sum_{i=1}^{r}\sum_{j=1}^{r} h_i h_j \bar{A}_{dlgij}, \ \bar{A}_{dlgij} = \begin{bmatrix} A_{di} & 0 \\ \hat{B}_{gj}\hat{S}_l C_{di} & \hat{A}_{gj} \end{bmatrix},$$

$$\bar{B}_{2dlgh} = \sum_{i=1}^{r}\sum_{j=1}^{r} h_i h_j \bar{B}_{2dlgij}, \ \bar{B}_{2dlgij} = \begin{bmatrix} B_{di} \\ \hat{B}_{gj}\hat{S}_l D_{di} \end{bmatrix},$$

$$\bar{B}_{1gh} = \sum_{i=1}^{r} h_i \bar{B}_{1gi}, \ \bar{B}_{1gi} = \begin{bmatrix} 0 \\ \hat{B}_{gi} \end{bmatrix}, \ \bar{E}_{dgh} = \sum_{i=1}^{r} h_i \bar{E}_{dgi},$$

$$\bar{E}_{dgi} = \begin{bmatrix} E_{di} & -\hat{E}_{gi} \end{bmatrix}, \ \bar{C}_{dh} = \sum_{i=1}^{r} h_i \bar{C}_{di}, \ \bar{C}_{di} = \begin{bmatrix} C_{di} & 0 \end{bmatrix},$$

$$\bar{D}_{dh} = \sum_{i=1}^{r} h_i D_{di}, \ \bar{F}_{dh} = \sum_{i=1}^{r} h_i F_{di}.$$

Remark 10.1 Note that although asynchronous modes of the sensor and filter are controlled by the plant mode directly, they are conditionally independent, namely,

$$\begin{aligned} &\Pr\{\varphi(t) = l, \nu(t) = g | r(t) = d\} \\ &= \Pr\{\varphi(t) = l | r(t) = d\} \times \Pr\{\nu(t) = g | r(t) = d\} \\ &= \varphi_{dl}\phi_{dg}. \end{aligned} \tag{10.13}$$

The main purpose of this chapter is to devise an asynchronous reliable filter (10.11), which meets the following two requirements:
(1) When $w(t) \equiv 0$, the filtering error system (10.12) is stochastically stable;
(2) Under the zero initial condition, the system (10.12) satisfies a given L_2–L_∞ performance γ ($\gamma > 0$), namely, for any $w(t) \in l_2[0, \infty)$, $e(t)$ satisfies

$$\sup_t \sqrt{E[||e(t)||^2]} < \gamma \sqrt{\int_0^\infty w^T(t)w(t)dt}. \tag{10.14}$$

10.3 Main Results

We start with analyzing the stochastic stability of the filtering error system (10.12) with a prescribed L_2–L_∞ performance and then develop an approach to compute filter gains.

Theorem 10.2 *For a given $\gamma > 0$, if there exist matrices $P_d > 0$, $R_{dh} = R_{dh}^T$, the diagonal matrix $W_l > 0$, and a scalar $\mu > 0$ for any $d \in \mathcal{Q}$, $l \in \mathcal{L}$ and $g \in \mathcal{G}$ such that following inequalities hold:*

$$\Upsilon^{dh} = \mathbf{Her}\left(\sum_{g=1}^{G}\sum_{l=1}^{L}\phi_{dg}\varphi_{dl}P_d\bar{A}_{dlgh}\right) + \sum_{e=1}^{Q}\pi_{de}P_e - R_{dh} < 0, \tag{10.15}$$

$$\Sigma^{gdh} = \begin{bmatrix} R_{dh} & P_d\bar{B}_{2dlgh} & P_d\bar{B}_{1gh}\check{S}_l & \bar{C}_{dh}^T W_l \\ * & -I & 0 & \bar{D}_{dh}^T W_l \\ * & * & -W_l & 0 \\ * & * & * & -W_l \end{bmatrix} < 0, \tag{10.16}$$

$$\Xi^{dh} = \begin{bmatrix} -\gamma^2 I & \Xi_{12}^{dh} & \Xi_{13}^{dh} \\ * & -P_d & 0 \\ * & * & -\mu I \end{bmatrix} < 0, \tag{10.17}$$

system (10.12) is stochastically stable with a prescribed L_2–L_∞ performance γ, where

$$\begin{aligned} \Xi_{12}^{dh} &= \left[\sqrt{\phi_{d1}}\bar{E}_{d1h}^T \cdots \sqrt{\phi_{dG}}\bar{E}_{dGh}^T\right]^T, \\ \Xi_{13}^{dh} &= \left[\sqrt{\phi_{d1}}\bar{F}_{dh}^T \cdots \sqrt{\phi_{dG}}\bar{F}_{dh}^T\right]^T, \\ \mathbf{Her}(A) &= A + A^T. \end{aligned}$$

Proof Due to the diagonal matrix $W_l > 0$, applying Schur Complement to (10.16), we have

$$\mathscr{R}_1^{dlgh} + \mathscr{R}_2^{dgh}\check{S}_l W_l^{-1}\check{S}_l(\mathscr{R}_2^{gdh})^T + \mathscr{R}_3^{dh}W_l(\mathscr{R}_3^{dh})^T < 0, \tag{10.18}$$

with

$$\mathscr{R}_1^{dlgh} = \begin{bmatrix} R_{dh} & P_d\bar{B}_{2dlgh} \\ * & -I \end{bmatrix}, \quad \mathscr{R}_2^{dgh} = \begin{bmatrix} P_d\bar{B}_{1gh} \\ 0 \end{bmatrix}, \quad \mathscr{R}_3^{dh} = \begin{bmatrix} \bar{C}_{dh}^T \\ \bar{D}_{dh}^T \end{bmatrix}.$$

From $\Delta_l(t)\Delta_l(t) \le \check{S}_l\check{S}_l \le I$, it follows that

$$\mathscr{R}_1^{dlgh} + \mathscr{R}_2^{gdh}\Delta_l(t)W_l^{-1}\Delta_l(t)(\mathscr{R}_2^{gdh})^T + \mathscr{R}_3^{dh}W_l(\mathscr{R}_3^{dh})^T < 0. \tag{10.19}$$

Furthermore, we obtain

$$\mathscr{R}_1^{dlgh} + \mathbf{Her}\left(\mathscr{R}_2^{gdh}\Delta_l(t)(\mathscr{R}_3^{dh})^T\right) < 0. \tag{10.20}$$

Associating (10.13), (10.15) and (10.20) with $\sum_{l=1}^{L}\varphi_{dl} = 1$ and $\sum_{g=1}^{G}\phi_{dg} = 1$, we have

$$\begin{bmatrix} \mathbf{Her}(\mathscr{A}_{dh}) + \sum_{e=1}^{Q}\pi_{de}P_e & \mathscr{B}_{dh} \\ * & -I \end{bmatrix} < 0, \tag{10.21}$$

where

$$\mathscr{A}_{dh} = \sum_{g=1}^{G}\sum_{l=1}^{L}\phi_{dg}\varphi_{dl}P_d\left(\bar{A}_{dlgh} + \bar{B}_{1gh}\Delta_l(t)\bar{C}_{dh}\right),$$

$$\mathscr{B}_{dh} = \sum_{g=1}^{G}\sum_{l=1}^{L}\phi_{dg}\varphi_{dl}P_d\left(\bar{B}_{2dlgh} + \bar{B}_{1gh}\Delta_l(t)\bar{D}_{dh}\right).$$

The candidate Lyapunov function for system (10.12) is chosen as

$$V(t) = \xi^T(t)P_{r(t)}\xi(t). \tag{10.22}$$

Define $V(t+\Delta t) = \xi^T(t+\Delta t)P_{r(t+\Delta t)}\xi(t+\Delta t)$ with $r(t+\Delta t) = e$. Let $\mathcal{A}$ be the weak infinitesimal generator of the stochastic process $\{\xi(t), r(t)\}$. It follows that

$$\mathcal{A}V(t) = \xi^T(t)\left(\sum_{e=1}^{Q}\pi_{de}P_e\right)\xi(t) + 2\dot{\xi}^T(t)P_d\xi(t). \tag{10.23}$$

Then taking the expectation, we obtain

$$E\{\mathcal{A}V(t)|r(t) = d\} = \zeta^T(t)\begin{bmatrix} \mathbf{Her}(\mathscr{A}_{dh}) + \sum_{e=1}^{Q}\pi_{de}P_e & \mathscr{B}_{dh} \\ * & 0 \end{bmatrix}\zeta(t), \tag{10.24}$$

where

$$\zeta(t) = [\xi^T(t)\ w^T(t)]^T. \tag{10.25}$$

Recalling (10.21) with $w(t) = 0$, $E\{\mathcal{A}V(t)|r(t) = d\} < 0$ holds, which ensures the stochastic stability of system (10.12).

On the other hand,

$$E\{e^T(t)e(t)\} = \zeta^T(t)\left(\sum_{g=1}^{G}\phi_{dg}\begin{bmatrix} \bar{E}_{dgh}^T\bar{E}_{dgh} & \bar{E}_{dgh}^T\bar{F}_{dh} \\ * & \bar{F}_{dh}^T\bar{F}_{dh} \end{bmatrix}\right)\zeta(t). \tag{10.26}$$

Applying Schur Complement to (10.17), we have that

$$\sum_{g=1}^{G} \phi_{dg} \begin{bmatrix} \bar{E}_{dgh}^T \bar{E}_{dgh} & \bar{E}_{dgh}^T \bar{F}_{dh} \\ * & \bar{F}_{dh}^T \bar{F}_{dh} \end{bmatrix} < \gamma^2 \begin{bmatrix} P_d & 0 \\ 0 & \mu I \end{bmatrix}. \tag{10.27}$$

In this case, we obtain that

$$E\{e^T(t)e(t)\} < \gamma^2 \xi^T(t) P_d \xi(t) + \gamma^2 \mu w^T(t) w(t). \tag{10.28}$$

Adopting the same method in [1], we can find a sufficiently small scalar parameter $\mu > 0$ to meet

$$\mu w^T(t) w(t) < \int_t^{\infty} w^T(t) w(t) \mathrm{d}t. \tag{10.29}$$

From (10.21) and (10.24), it follows that $E\{\mathcal{A}V(t)\} < w^T(t)w(t)$. Under the zero initial condition, we obtain

$$V(t) = \xi^T(t) P_d \xi(t) < \int_0^t w^T(t) w(t) \mathrm{d}t. \tag{10.30}$$

According to (10.28)–(10.30), it follows that

$$E\{e^T(t)e(t)\} < \gamma^2 \int_0^{\infty} w^T(t) w(t) \mathrm{d}t. \tag{10.31}$$

Considering (10.14), we can conclude that system (10.12) satisfies the L_2–L_∞ performance index. The proof is completed.

Based on Theorem 10.2 and the Lyapunov function matrix, the following theorem show a solution to the filtering gains in terms of LMIs.

Theorem 10.3 *For a given $\gamma > 0$, if there exist matrices $P_{1d} > 0$, $P_2 > 0$, the symmetric matrix $\begin{bmatrix} R_{1dij} & R_{2dij} \\ * & R_{3dij} \end{bmatrix}$, the diagonal matrix $W_l > 0$, $\mathcal{A}_{gi}$, $\mathcal{B}_{gi}$, $\mathcal{E}_{gi}$ and a scalar $\mu > 0$ for any $d \in \mathcal{Q}$, $l \in \mathcal{L}$, $g \in \mathcal{G}$ and $i, j \in \mathcal{I}$ such that following inequalities hold:*

$$\Upsilon^{dii} < 0, \tag{10.32}$$

$$\Upsilon^{dij} + \Upsilon^{dji} < 0, \ i < j, \tag{10.33}$$

$$\Sigma^{dlgii} < 0, \tag{10.34}$$

$$\Sigma^{dlgij} + \Sigma^{dlgji} < 0, \ i < j, \tag{10.35}$$

$$\Xi^{di} = \begin{bmatrix} -\gamma^2 I & \Xi_{12}^{di} & \Xi_{13}^{di} & \Xi_{14}^{di} \\ * & -P_{1d} & -P_2 & 0 \\ * & * & -P_2 & 0 \\ * & * & * & -\mu I \end{bmatrix} < 0, \tag{10.36}$$

system (10.12) is stochastically stable with a prescribed L_2–L_∞ performance. The desired filter parameters can be derived from

$$\hat{A}_{gi} = P_2^{-1}\mathcal{A}_{gi}, \ \hat{B}_{gi} = P_2^{-1}\mathcal{B}_{gi}, \hat{E}_{gi} = \mathcal{E}_{gi}, \tag{10.37}$$

where

$$\Upsilon^{dij} = \begin{bmatrix} \Upsilon_{11}^{dij} - R_{1dij} & \Upsilon_{12}^{dij} - R_{2dij} \\ * & \mathbf{Her}(\sum_{g=1}^{G}\phi_{dg}\mathcal{A}_{gj}) - R_{3dij} \end{bmatrix},$$

$$\Sigma_{dlgij} = \begin{bmatrix} R_{1dij} & R_{2dij} & \Sigma_{13}^{dlgij} & \mathcal{B}_{gj}\check{S}_l & C_{di}^T W_l \\ * & R_{3dij} & \Sigma_{23}^{dlgij} & \mathcal{B}_{gj}\check{S}_l & 0 \\ * & * & -I & 0 & D_{di}^T W_l \\ * & * & * & -W_l & 0 \\ * & * & * & * & -W_l \end{bmatrix},$$

$$\Upsilon_{11}^{dij} = \mathbf{Her}\left(\sum_{g=1}^{G}\sum_{l=1}^{L}\phi_{dg}\varphi_{dl}(P_{1d}A_{di} + \mathcal{B}_{gj}\hat{S}_l C_{di})\right) + \sum_{e=1}^{Q}\pi_{de}P_{1e},$$

$$\Upsilon_{12}^{dij} = \sum_{g=1}^{G}\sum_{l=1}^{L}\phi_{dg}\varphi_{dl}(\mathcal{B}_{gj}\hat{S}_l C_{di})^T + \sum_{g=1}^{G}\phi_{dg}\mathcal{A}_{gj} + A_{di}^T P_2,$$

$$\Sigma_{13}^{dlgij} = P_{1d}B_{di} + \mathcal{B}_{gj}\hat{S}_l D_{di}, \ \Sigma_{23}^{dlgij} = P_2 B_{di} + \mathcal{B}_{gj}\hat{S}_l D_{di},$$

$$\Xi_{12}^{di} = \left[\sqrt{\phi_{d1}}E_{di}^T \dots \sqrt{\phi_{dG}}E_{di}^T\right]^T, \ \Xi_{13}^{di} = -\left[\sqrt{\phi_{d1}}\mathcal{E}_{1i}^T \dots \sqrt{\phi_{dG}}\mathcal{E}_{Gi}^T\right]^T,$$

$$\Xi_{14}^{di} = \left[\sqrt{\phi_{d1}}F_{di}^T \dots \sqrt{\phi_{dG}}F_{di}^T\right]^T.$$

Proof The Lyapunov function matrices P_d and R_{dh} are defined as

$$P_d = \begin{bmatrix} P_{1d} & P_2 \\ * & P_2 \end{bmatrix}, \ R_{dh} = \sum_{i=1}^{r}\sum_{j=1}^{r} h_i h_j \begin{bmatrix} R_{1dij} & R_{2dij} \\ * & R_{3dij} \end{bmatrix}. \tag{10.38}$$

Then based on the fuzzy inference, from (10.12) and (10.37), we have

$$
\begin{aligned}
\Upsilon^{dh} &= \sum_{i=1}^{r}\sum_{j=1}^{r} h_i h_j \Upsilon^{dij} \\
&= \sum_{i=1}^{r} h_i^2 \Upsilon^{dii} + \sum_{i=1}^{r-1}\sum_{j=i+1}^{r} (h_i h_j \Upsilon^{dij} + h_j h_i \Upsilon^{dji}), \\
\Sigma^{dlgh} &= \sum_{i=1}^{r}\sum_{j=1}^{r} h_i h_j \Sigma^{dlgij} \\
&= \sum_{i=1}^{r} h_i^2 \Sigma^{dlgii} + \sum_{i=1}^{r-1}\sum_{j=i+1}^{r} (h_i h_j \Sigma^{dlgij} + h_j h_i \Sigma^{dlgji}), \\
\Xi^{dh} &= \sum_{i=1}^{r} h_i \Xi^{di}.
\end{aligned} \tag{10.39}
$$

Hence, we clearly observe that: the validity of (10.15) is ensured by (10.32) and (10.33); (10.34) and (10.35) can guarantee the correctness of (10.16); (10.36) is equivalent to (10.17). Furthermore, if LMIs (10.32)–(10.36) are feasible, L_2–L_∞ filter gains can be derived by (10.37). The proof is completed.

Remark 10.4 We employ the slack symmetric matrix R_{dij} and the stochastic Lyapunov function matrix $P_d = \begin{bmatrix} P_{1d} & P_2 \\ * & P_3 \end{bmatrix}$ with a certain structure $P_2 = P_3$ to solve filtering gains, which change a complex nonlinear problem to a liner one, as shown in Theorem 10.3. Besides, we can obtain the optimal filter with the smallest L_2–L_∞ performance γ by solving the following convex optimization issue:

$$
\begin{aligned}
&\min \quad \rho \\
&\text{subject to (10.32)} - -\text{(10.36) with } \rho = \gamma^2.
\end{aligned}
$$

The corresponding minimal value of γ can be inferred as $\gamma_{min} = \sqrt{\rho}$.

10.4 Illustrative Example

In this section, we provide an example to illustrate the correctness and applicability of our proposed technique. System parameters are given as

$$
A_{11} = \begin{bmatrix} -0.2 & 40 \\ -1 & -8 \end{bmatrix}, \; A_{12} = \begin{bmatrix} -3.6 & 45 \\ -2 & -7 \end{bmatrix},
$$

$$
A_{21} = \begin{bmatrix} -0.1 & 46 \\ -1 & -10 \end{bmatrix}, \; A_{22} = \begin{bmatrix} -4.0 & 50 \\ -1.2 & -9 \end{bmatrix},
$$

$$
\begin{aligned}
&C_{11} = \begin{bmatrix} 1.0 & 0 \\ 0 & 1.0 \end{bmatrix}, \; C_{12} = \begin{bmatrix} 1 & 0 \\ 0.1 & 0.9 \end{bmatrix}, \\
&C_{21} = \begin{bmatrix} 1 & 0 \\ -0.1 & 0.6 \end{bmatrix}, \; C_{22} = \begin{bmatrix} 1 & 0 \\ 0.1 & 0.7 \end{bmatrix}, \\
&F_{11} = 0.1, \; F_{12} = -0.1, \; F_{21} = 0.2, \; F_{22} = 0.3, \\
&B_{di} = \begin{bmatrix} 0 \\ 1 \end{bmatrix}, \; D_{di} = \begin{bmatrix} 1 & 0.1 \end{bmatrix}, \; E_{di} = \begin{bmatrix} 1 & 0 \end{bmatrix}, \\
&h_1 = 0.5(\sin^2(w(k))), \; h_2 = 1 - h_1, \; d, \; i \in \{1, 2\}.
\end{aligned}
$$

We assume that 2 sensors operate in this example and each has 3 jump modes, subject to

$$
\Psi = \begin{bmatrix} 0.2 & 0.3 & 0.5 \\ 0.4 & 0.5 & 0.1 \end{bmatrix},
$$

and sensor failures are supposed to be

$$
\begin{aligned}
&\hat{S}_1 = \text{diag}\{0.8, \; 0.9\}, \; \check{S}_1 = \text{diag}\{0.05, \; 0.05\}, \\
&\hat{S}_2 = \text{diag}\{0.9, \; 1.0\}, \; \check{S}_2 = \text{diag}\{0.1, \; 0\}, \\
&\hat{S}_3 = \text{diag}\{1.0, \; 0.9\}, \; \check{S}_3 = \text{diag}\{0, \; 0.1\}.
\end{aligned}
$$

$\Delta_l(t)$ $(l \in \{1, 2, 3\})$ is set as

$$
\begin{aligned}
&\Delta_1(t) = \text{diag}\{0.05\sin(t), \; 0.05\cos(t)\}, \\
&\Delta_2(t) = \text{diag}\{0.1\sin(t), \; 0\}, \; \Delta_3(t) = \text{diag}\{0, \; 0.1\cos(t)\}.
\end{aligned}
$$

The original system and designed filter work by obeying the following probability matrices:

$$
\Pi = \begin{bmatrix} -6 & 6 \\ 7 & -7 \end{bmatrix}, \; \Phi = \begin{bmatrix} 0.3 & 0.7 \\ 0.6 & 0.4 \end{bmatrix}.
$$

By solving LMIs in Theorem 10.3, we obtain the minimal $\gamma^* = 0.2659$ and the following filtering parameters:

$$
\begin{aligned}
&\hat{A}_{11} = \begin{bmatrix} -0.0411 & 53.8840 \\ -1.8045 & -12.6273 \end{bmatrix}, \; \hat{B}_{11} = \begin{bmatrix} -0.0401 & 0.0106 \\ -1.0890 & -0.2445 \end{bmatrix}, \\
&\hat{A}_{12} = \begin{bmatrix} -4.9147 & 55.9648 \\ -1.0493 & -11.9975 \end{bmatrix}, \; \hat{B}_{12} = \begin{bmatrix} 0.0847 & 0.1908 \\ -1.0392 & -0.5728 \end{bmatrix}, \\
&\hat{A}_{21} = \begin{bmatrix} -0.3436 & 33.4982 \\ -2.1086 & -6.2658 \end{bmatrix}, \; \hat{B}_{21} = \begin{bmatrix} -0.0405 & 0.0144 \\ -1.0889 & -0.2461 \end{bmatrix},
\end{aligned}
$$

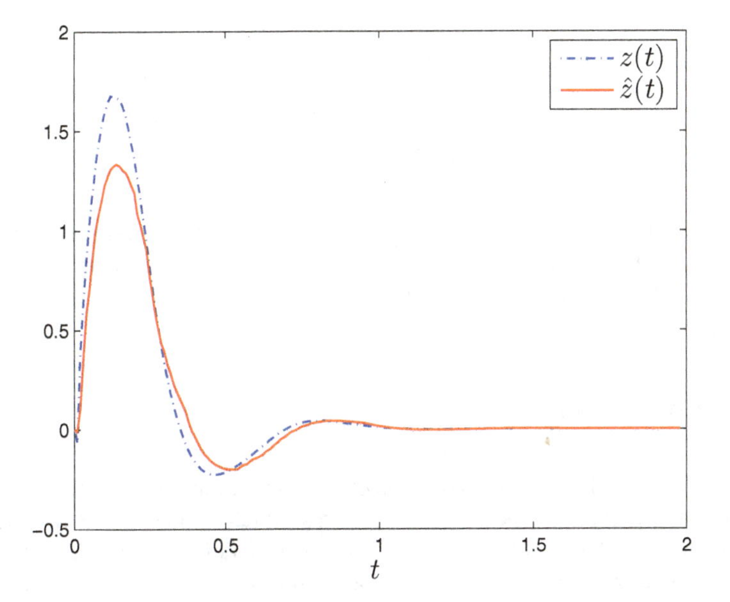

Fig. 10.1 Output $z(t)$ and its estimation $\hat{z}(t)$)

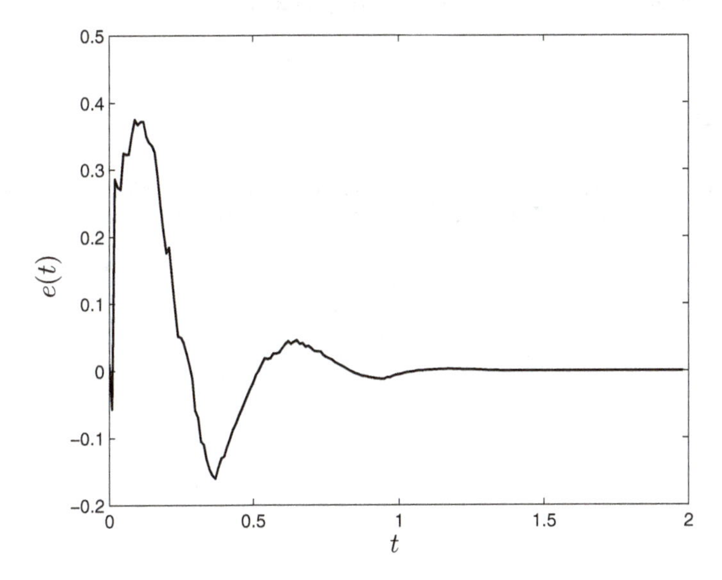

Fig. 10.2 Estimation error $e(t)$

$$\hat{A}_{22} = \begin{bmatrix} -2.2730 & 42.4311 \\ -3.8172 & -5.3119 \end{bmatrix}, \ \hat{B}_{22} = \begin{bmatrix} 0.0847 & 0.1908 \\ -1.0392 & -0.5729 \end{bmatrix},$$
$$\hat{E}_{11} = \begin{bmatrix} -0.9944 & 0.0238 \end{bmatrix}, \ \hat{E}_{12} = \begin{bmatrix} -0.9944 & 0.0238 \end{bmatrix},$$
$$\hat{E}_{21} = \begin{bmatrix} -0.9937 & 0.0092 \end{bmatrix}, \ \hat{E}_{22} = \begin{bmatrix} -0.9937 & 0.0092 \end{bmatrix}.$$

With initial state $x(0) = \begin{bmatrix} 0.1 & 0.2 \end{bmatrix}^T$ and noise $w(t) = 100e^{-t}\sin(t)$, we can clearly observe from Figs. 10.1, 10.2 that $\hat{z}(t)$ can track $z(t)$, and $e(t)$ approaches to zero over time, which imply that our method is correct and applicable.

10.5 Conclusion

In this chapter, we have addressed the asynchronous reliable L_2–L_∞ filtering design issue for the continuous-time T–S fuzzy systems, which is subject to stochastic sensor failures. An LMI-based sufficient condition has been developed for the existence of the non-synchronous filter and ensuring that the filtering error system is stochastically stable with a given L_2–L_∞ performance. An example has been presented to verify the validity and feasibility of the proposed approach.

Reference

1. Chang, X.-H., Park, J.H., Shi, P.: Fuzzy resilient energy-to-peak filtering for continuous-time nonlinear systems. IEEE Trans. Fuzzy Syst. **25**(6), 1576–1588 (2017)

Chapter 11
HMM-Based Asynchronous Filter Design of Continuous-Time Fuzzy MJSs

11.1 Introduction

In this chapter, we focus on the problem of dissipative asynchronous filtering for nonlinear continuous-time MJSs, described by the T–S fuzzy technique. A dissipative asynchronous fuzzy filter is constructed by the HMM. We utilize the stochastic Lyapunov technique and introduce some slack matrices to obtain a sufficient condition guaranteeing the stochastic stability of the filtering error systems with dissipative performance. Then based on this sufficient condition and Finsler's lemma, two filtering approaches are developed in terms of LMIs.

11.2 Preliminary Analysis

Consider the following continuous-time MJSs, modeled by the T–S fuzzy approach:
Plant rule i: IF $\theta_1(t)$ is μ_{i1}, $\theta_2(t)$ is μ_{i2}, ..., and $\theta_p(t)$ is μ_{ip}, THEN

$$\begin{cases} \dot{x}(t) = A_{\delta(t)i}x(t) + B_{\delta(t)i}w(t), \\ y(t) = C_{\delta(t)i}x(t) + D_{\delta(t)i}w(t), \\ z(t) = E_{\delta(t)i}x(t) + F_{\delta(t)i}w(t), \end{cases} \tag{11.1}$$

where $\theta_j(t)$ $(j \in \{1, 2, \ldots, p\})$ is the premise variable and μ_{ij} $(i \in \mathcal{I} = \{1, 2, \ldots, r\})$ is the fuzzy set with r fuzzy rules. $x(t) \in R^{n_x}$ is the state; $y(t) \in R^{n_y}$ is the measured output; $z(t) \in R^{n_z}$ is the signal to be estimated; and $w(t) \in R^{n_w}$ is the disturbance, belonging to $l_2[0, +\infty)$. The parameter $\delta(t)$ is applied to describe a continuous-time Markov jump process with right continuous trajectories, and it takes a value in a finite set $\mathcal{M} = \{1, 2, \ldots, M\}$ by obeying the transition rate matrix $\Phi = [\vartheta_{mn}]$ and

S. Dong et al., *Control and Filtering of Fuzzy Systems with Switched Parameters*, Studies in Systems, Decision and Control 268,
https://doi.org/10.1007/978-3-030-35566-1_11

$$\Pr\{\delta(t+\Delta t)=n|\delta(t)=m\}=\begin{cases}\vartheta_{mn}\Delta t+o(\Delta t), & n\neq m,\\ 1+\vartheta_{mm}\Delta t+o(\Delta t), n=m,\end{cases}\tag{11.2}$$

where Δt ($\Delta t>0$) is the the infinitesimal transition time interval, implying that $\lim_{\Delta t\to 0}\frac{o(\Delta t)}{\Delta t}=0$. The variable ϑ_{mn} ($\vartheta_{mn}\geq 0,\ m\neq n$) is the transition rate from mode m at time t to mode n at time $t+\Delta t$, satisfying $\vartheta_{mm}=-\sum_{n=1,n\neq m}^{M}\vartheta_{mn}$.

Using the T–S inference method with $\delta(t)=m$, the overall system can be rewritten as

$$\begin{cases}\dot{x}(t)=&A_{mh}x(t)+B_{mh}w(t),\\ y(t)=&C_{mh}x(t)+D_{mh}w(t),\\ z(t)=&E_{mh}x(t)+F_{mh}w(t),\end{cases}\tag{11.3}$$

where

$$A_{mh}=\sum_{i=1}^{r}h_i(\theta(t))A_{mi},\ B_{mh}=\sum_{i=1}^{r}h_i(\theta(t))B_{mi},$$
$$C_{mh}=\sum_{i=1}^{r}h_i(\theta(t))C_{mi},\ D_{mh}=\sum_{i=1}^{r}h_i(\theta(t))D_{mi},$$
$$E_{mh}=\sum_{i=1}^{r}h_i(\theta(t))E_{mi},\ F_{mh}=\sum_{i=1}^{r}h_i(\theta(t))F_{mi},$$
$$\theta(t)=[\theta_1(t),\theta_1(t),\ldots,\theta_p(t)],\ h_i(\theta(t))=\frac{\prod_{j=1}^{p}\eta_{ij}(\theta_j(t))}{\sum_{i=1}^{r}\prod_{j=1}^{p}\mu_{ij}(\theta_j(t))}.$$

For simplification, denote $h_i(\theta(t))$ as h_i.

In this chapter, we design an asynchronous fuzzy filter by the PDC approach, and the asynchronous phenomenon between the original plant and the filter is described by the HMM. Construct the following asynchronous filter:

$$\begin{cases}\dot{\hat{x}}(t)=&A_{f\rho(t)h}\hat{x}(t)+B_{f\rho(t)h}y(t),\\ \hat{z}(t)=&E_{f\rho(t)h}\hat{x}(t)+F_{f\rho(t)h}y(t),\end{cases}\tag{11.4}$$

where

$$A_{f\rho(t)h}=\sum_{i=1}^{r}h_iA_{f\rho(t)i},\ B_{f\rho(t)h}=\sum_{i=1}^{r}h_iB_{f\rho(t)i},$$
$$E_{f\rho(t)h}=\sum_{i=1}^{r}h_iE_{f\rho(t)i},\ F_{f\rho(t)h}=\sum_{i=1}^{r}h_iF_{f\rho(t)i}.$$

Variable $\rho(t)$ ($\rho(t)\in\mathcal{S}=\{1,2,\ldots,S\}$) is introduced to describe the non-synchronization, and satisfies the conditional probability matrix $\Psi=[\pi_{ms}]$ with

$$\Pr\{\rho(t) = s|\delta(t) = m\} = \pi_{ms}, \quad \sum_{s=1}^{S} \pi_{ms} = 1, \tag{11.5}$$

where $\pi_{ms} \geq 0$.

Define $\xi(t) = \left[x^T(t)\ \hat{x}^T(t)\right]^T$ and $e(t) = z(t) - \hat{z}(t)$. Then, based on (11.3) and (11.4) with $\rho(t) = s$, we obtain the following filtering error system:

$$\begin{cases} \dot{\xi}(t) = \bar{A}_{msh}\xi(t) + \bar{B}_{msh}w(t), \\ e(t) = \bar{E}_{msh}\xi(t) + \bar{F}_{msh}w(t), \end{cases} \tag{11.6}$$

where

$$\bar{A}_{msh} = \sum_{i=1}^{r}\sum_{j=1}^{r} h_i h_j \bar{A}_{msij}, \quad \bar{B}_{msh} = \sum_{i=1}^{r}\sum_{j=1}^{r} h_i h_j \bar{B}_{msij},$$

$$\bar{E}_{msh} = \sum_{i=1}^{r}\sum_{j=1}^{r} h_i h_j \bar{E}_{msij}, \quad \bar{F}_{msh} = \sum_{i=1}^{r}\sum_{j=1}^{r} h_i h_j \bar{F}_{msij},$$

$$\bar{A}_{msij} = \begin{bmatrix} A_{mi} & 0 \\ B_{fsj}C_{mi} & A_{fsj} \end{bmatrix}, \quad \bar{B}_{msij} = \begin{bmatrix} B_{mi} \\ B_{fsj}D_{mi} \end{bmatrix},$$

$$\bar{E}_{msij} = \left[E_{mi} - F_{fsj}C_{mi}\ \ -E_{fsj}\right], \quad \bar{F}_{msij} = F_{mi} - F_{fsj}D_{mi}.$$

In this chapter, our objective is to design a dissipative asynchronous filter, which satisfies the following two conditions:
(1) When $w(t) \equiv 0$, system (11.6) is stochastically stable;
(2) Under the zero initial condition, system (11.6) is strictly $(\mathcal{U},\ \mathcal{G},\ \mathcal{V})$-$\alpha$-dissipative, that is,

$$J(e(t), w(t), \mathcal{T}) > \alpha \int_0^{\mathcal{T}} w^T(t)w(t)\mathrm{d}t, \tag{11.7}$$

where

$$J(e(t), w(t), \mathcal{T}) = E\left\{\int_0^{\mathcal{T}} e^T(t)\mathcal{U}e(t) + 2e^T(t)\mathcal{G}w(t) + w^T(t)\mathcal{V}w(t)\mathrm{d}t\right\},$$

$$\mathcal{U} \leq 0,\ \mathcal{V}^T = \mathcal{V},\ \alpha > 0.$$

11.3 Main Results

In this section, we firstly analyze the stochastic stability and strict dissipativity of the filtering error system in (11.6). Then a filtering design approach for the asynchronous dissipative filtering issue is developed.

Theorem 11.1 *If there exist matrices $P_m > 0$, $\bar{P}_m = \sum_{n=1}^{M} \vartheta_{mn} P_n$, R_s, G_s, H and symmetric matrix W_{msi} for any $i,\ j \in \mathcal{I}$, $m \in \mathcal{M}$ and $s \in \mathcal{S}$ satisfying*

$$\sum_{s=1}^{S} \pi_{ms} W_{msi} < \mathcal{V} - \alpha I, \tag{11.8}$$

$$\Lambda_{msii} < 0, \tag{11.9}$$

$$\Lambda_{msij} + \Lambda_{msji} < 0,\ (i < j). \tag{11.10}$$

where

$$\Lambda_{msij} = \begin{bmatrix} \Lambda_s^{11} & \Lambda_{msij}^{12} & R_s \bar{B}_{msij} & 0 \\ * & \Lambda_{msij}^{22} & G_s \bar{B}_{msij} & \bar{E}_{msij}^T H^T \\ * & * & -W_{msi} & \bar{F}_{msij}^T H^T - \mathcal{G}^T \\ * & * & * & -\mathcal{U} - \mathbf{Her}(H) \end{bmatrix},$$
$$\Lambda_s^{11} = -R_s - R_s^T,\ \Lambda_{msij}^{12} = P_m + R_s \bar{A}_{msij} - G_s^T,$$
$$\Lambda_{msij}^{22} = \bar{P}_m + \mathbf{Her}(G_s \bar{A}_{msij}),\ \mathbf{Her}(X) = X + X^T,$$

the filtering error system (11.6) is stochastically stable with strict ($\mathcal{U}$, $\mathcal{G}$, $\mathcal{V}$)-α dissipativity.

Proof Define

$$W_{msh} = \sum_{i=1}^{r} h_i W_{msi}. \tag{11.11}$$

According to the T–S fuzzy principle, (11.6) and (11.8)–(11.10), we have

$$\sum_{s=1}^{S} \pi_{ms} W_{msh} < \mathcal{V} - \alpha I, \tag{11.12}$$

and

$$\begin{aligned} \Lambda_{msh} &= \sum_{i=1}^{r} \sum_{j=1}^{r} h_i h_j \Lambda_{msij} \\ &= \sum_{i=1}^{r} h_i^2 \Lambda_{msii} + \sum_{i=1}^{r-1} \sum_{j=i+1}^{r} h_i h_j (\Lambda_{msij} + \Lambda_{msji}) < 0, \end{aligned} \tag{11.13}$$

with

$$\Lambda_{msh} = \begin{bmatrix} \Lambda_s^{11} & \Lambda_{msh}^{12} & R_s \bar{B}_{msh} & 0 \\ * & \Lambda_{msh}^{22} & G_s \bar{B}_{msh} & \bar{E}_{msh}^T H^T \\ * & * & -W_{msh} & \bar{F}_{msh}^T H^T - \mathcal{G}^T \\ * & * & * & -\mathcal{U} - \mathbf{Her}(H) \end{bmatrix},$$

$$\Lambda_{msh}^{12} = P_m + R_s \bar{A}_{msh} - G_s^T, \ \Lambda_{msh}^{22} = \bar{P}_m + \mathbf{Her}(G_s \bar{A}_{msh}).$$

Considering (11.12) with $\pi_{ms} \geq 0$, we can easily obtain

$$\sum_{s=1}^{S} \pi_{ms} \Lambda_{msh}^{'} < 0, \tag{11.14}$$

where

$$\Lambda_{msh}^{'} = \begin{bmatrix} \Lambda_s^{11} & \Lambda_{msh}^{12} & R_s \bar{B}_{msh} & 0 \\ * & \Lambda_{msh}^{22} & G_s \bar{B}_{msh} & \bar{E}_{msh}^T H^T \\ * & * & -\mathcal{V} + \alpha I & \bar{F}_{msh}^T H^T - \mathcal{G}^T \\ * & * & * & -\mathcal{U} - \mathbf{Her}(H) \end{bmatrix}.$$

It can be derived readily that

$$\Omega_m^1 + \mathbf{Her}\left(\sum_{s=1}^{S} \pi_{ms} \Upsilon_s^1 \left[-I \ \bar{A}_{msh} \right] \right) < 0, \tag{11.15}$$

and

$$\Omega_{msh}^2 + \mathbf{Her}\left(\sum_{s=1}^{S} \pi_{ms} \Upsilon_s^2 \begin{bmatrix} -I & \bar{A}_{msh} & \bar{B}_{msh} & 0 \\ 0 & \bar{E}_{msh} & \bar{F}_{msh} & -I \end{bmatrix} \right) < 0 \tag{11.16}$$

hold with

$$\Omega_m^1 = \begin{bmatrix} 0 & P_m \\ * & \bar{P}_m \end{bmatrix}, \ \Upsilon_s^1 = \begin{bmatrix} R_s \\ G_s \end{bmatrix},$$

$$\Omega_m^2 = \begin{bmatrix} 0 & P_m & 0 & 0 \\ * & \bar{P}_m & 0 & 0 \\ * & * & -\mathcal{V} + \alpha I & -\mathcal{G}^T \\ * & * & * & -\mathcal{U} \end{bmatrix}, \ \Upsilon_s^2 = \begin{bmatrix} R_s & 0 \\ G_s & 0 \\ 0 & 0 \\ 0 & H \end{bmatrix}.$$

Construct the following Lyapunov function for system (11.6):

$$V(t) = \xi^T(t) P_{\delta(t)} \xi(t). \tag{11.17}$$

It is supposed that $\delta(t) = m$ and $\delta(t + \Delta t) = n$. Let $\mathcal{A}$ be the weak infinitesimal generator of the stochastic process $\{\xi(t), \delta(t)\}$. We obtain

$$\mathcal{A}V(t) = 2\dot{\xi}^T(t)P_m\xi(t) + \xi^T(t)\Bigg(\sum_{n=1}^{M}\vartheta_{mn}P_n\Bigg)\xi(t). \tag{11.18}$$

Now, we introduce slack matrices R_s and G_s. From (11.6) with $w(t) = 0$, it is clearly obtained that

$$\Upsilon_s^1\left[-I\ \bar{A}_{msh}\right]\begin{bmatrix}\dot{\xi}(t)\\ \xi(t)\end{bmatrix} = 0. \tag{11.19}$$

Associating with (11.18) and $\pi_{ms} \geq 0$, and then taking expectation, we have

$$E\{\mathcal{A}V(t)|\delta(t) = m\} = E\left\{\begin{bmatrix}\dot{\xi}(t)\\ \xi(t)\end{bmatrix}^T \psi_{msh}^1\begin{bmatrix}\dot{\xi}(t)\\ \xi(t)\end{bmatrix}\right\}, \tag{11.20}$$

where

$$\psi_{msh}^1 = \Omega_m^1 + \mathbf{Her}\Bigg(\sum_{s=1}^{S}\pi_{ms}\Upsilon_s^1\left[-I\ \bar{A}_{msh}\right]\Bigg).$$

According to (11.15), we can conclude clearly that $E\{\mathcal{A}V(t)|\delta(t) = m\} < 0$, implying that the filtering error system (11.6) is stochastically stable.

Next, we focus on analyzing the dissipative performance of the system in (11.6).

Another slack matrix H is introduced. Together with R_s and G_s, when $w(t) \neq 0$, it follows from (11.6) that

$$\Upsilon_s^2\begin{bmatrix}-I & \bar{A}_{msh} & \bar{B}_{msh} & 0\\ 0 & \bar{E}_{msh} & \bar{F}_{msh} & -I\end{bmatrix}\begin{bmatrix}\dot{\xi}(t)\\ \xi(t)\\ w(t)\\ e(t)\end{bmatrix} = 0. \tag{11.21}$$

Then, we can derive that

$$\begin{aligned}&E\{\mathcal{A}V(t) - S(e(t), w(t), \mathcal{T}) + \alpha w^T(t)w(t)\}\\ &= E\{\zeta^T(t)\psi_{msh}^2\zeta(t)\},\end{aligned} \tag{11.22}$$

where

$$\zeta^T(t) = \left[\dot{\xi}^T(t)\ \xi^T(t)\ w^T(t)\ e^T(t)\right],$$

$$S(e(t), w(t), \mathcal{T}) = e^T(t)\mathcal{U}e(t) + 2e^T(t)\mathcal{G}w(t) + w^T(t)\mathcal{V}w(t),$$

$$\psi_{msh}^2 = \Omega_{msh}^2 + \mathbf{Her}\Bigg(\sum_{s=1}^{S}\pi_{ms}\Upsilon_s^2\begin{bmatrix}-I & \bar{A}_{msh} & \bar{B}_{msh} & 0\\ 0 & \bar{E}_{msh} & \bar{F}_{msh} & -I\end{bmatrix}\Bigg).$$

Based on (11.16), we have

$$E\{\mathcal{A}V(t) - S(e(t), w(t), \mathcal{T}) + \alpha w^T(t)w(t)\} < 0. \tag{11.23}$$

Using the integral computation with the zero initial condition, we have

$$\int_0^T E\{S(z(t), w(t))\}dt > E\{V(T)\} + \alpha \int_0^T w^T(t)w(t)dt. \tag{11.24}$$

According to (11.7), we can conclude that system (11.6) is strictly dissipative. The proof is completed.

Then based on Theorem 11.1, the asynchronous filtering approach is proposed by using the convex technique and selecting the appropriate Lyapunov matrix.

Theorem 11.2 *If there exist matrices* $P_m = \begin{bmatrix} P_{1m} & P_{2m} \\ * & P_{3m} \end{bmatrix} > 0$, R_{1s}, R_{2s} G_{1s}, G_{2s}, X_s, H *and symmetric matrix* W_{msi} *for any* $i,\ j \in \mathcal{I}$, $m \in \mathcal{M}$ *and* $s \in \mathcal{S}$ *satisfying*

$$\sum_{s=1}^{S} \pi_{ms} W_{msi} < \mathcal{V} - \alpha I, \tag{11.25}$$

$$\Gamma_{msii} < 0, \tag{11.26}$$

$$\Gamma_{msij} + \Gamma_{msji} < 0,\ (i < j), \tag{11.27}$$

where

$$\Gamma_{msij} = \begin{bmatrix} \Gamma_s^{11} & \Gamma_s^{12} & \Gamma_{msij}^{13} & \Gamma_{msij}^{14} & \Gamma_{msij}^{15} & 0 \\ * & \Gamma_s^{22} & \Gamma_{msij}^{23} & \Gamma_{msij}^{24} & \Gamma_{msij}^{25} & 0 \\ * & * & \Gamma_{msij}^{33} & \Gamma_{msij}^{34} & \Gamma_{msij}^{35} & \Gamma_{msij}^{36} \\ * & * & * & \Gamma_{msij}^{44} & \Gamma_{msij}^{45} & \Gamma_{msij}^{46} \\ * & * & * & * & -W_{msi} & \Gamma_{msij}^{56} \\ * & * & * & * & * & \Gamma_{msij}^{66} \end{bmatrix},$$

$\Gamma_s^{11} = -\mathbf{Her}(R_{1s}),\ \Gamma_s^{12} = -X_s - R_{2s}^T,\ \Gamma_s^{22} = -\mathbf{Her}(X_s),$
$\Gamma_{msij}^{13} = P_{1m} + R_{1s}A_{mi} + \mathcal{B}_{fsj}C_{mi} - G_{1s}^T,$
$\Gamma_{msij}^{23} = P_{2m}^T + R_{2s}A_{mi} + \mathcal{B}_{fsj}C_{mi} - X_s^T,$
$\Gamma_{msij}^{33} = \bar{P}_{1m} + \mathbf{Her}(G_{1s}A_{mi} + \mathcal{B}_{fsj}C_{mi}),$
$\Gamma_{msij}^{14} = P_{2m} + \mathcal{A}_{fsj} - G_{2s}^T,\ \Gamma_{msij}^{24} = P_{3m} + \mathcal{A}_{fsj} - X_s^T,$
$\Gamma_{msij}^{34} = \bar{P}_{2m} + \mathcal{A}_{fsj} + (G_{2s}A_{mi} + \mathcal{B}_{fsj}C_{mi})^T,$
$\Gamma_{msij}^{44} = \bar{P}_{3m} + \mathbf{Her}(\mathcal{A}_{fsj}),\ \Gamma_{msij}^{15} = R_{1s}B_{mi} + \mathcal{B}_{fsj}D_{mi},$
$\Gamma_{msij}^{25} = R_{2s}B_{mi} + \mathcal{B}_{fsj}D_{mi},\ \Gamma_{msij}^{66} = -\mathcal{U} - \mathbf{Her}(H),$
$\Gamma_{msij}^{35} = G_{1s}B_{mi} + \mathcal{B}_{fsj}D_{mi},\ \Gamma_{msij}^{45} = G_{2s}B_{mi} + \mathcal{B}_{fsj}D_{mi},$

$$\Gamma_{msij}^{46} = -\mathcal{E}_{fsj}^T H^T, \ \Gamma_{msij}^{56} = F_{mi}^T H^T - D_{mi}^T \mathcal{F}_{fsj}^T H^T - \mathcal{G}^T,$$

$$\Gamma_{msij}^{36} = E_{mi}^T H^T - C_{mi}^T \mathcal{F}_{fsj}^T H^T, \ \begin{bmatrix} \bar{P}_{1m} & \bar{P}_{2m} \\ * & \bar{P}_{3m} \end{bmatrix} = \sum_{n=1}^{M} \vartheta_{mn} \begin{bmatrix} P_{1n} & P_{2n} \\ * & P_{3n} \end{bmatrix},$$

the filtering error system (11.6) is stochastically stable with strict ($\mathcal{U}$, $\mathcal{G}$, $\mathcal{V}$)-α dissipativity. Furthermore, we can derive the asynchronous dissipative fuzzy filtering parameters from

$$\begin{aligned} A_{fsi} &= X_s^{-1}\mathcal{A}_{fsi}, \ E_{fsi} = \mathcal{E}_{fsi}, \\ B_{fsi} &= X_s^{-1}\mathcal{B}_{fsi}, \ F_{fsi} = \mathcal{F}_{fsi}. \end{aligned} \tag{11.28}$$

Proof Define

$$\begin{aligned} P_m &= \begin{bmatrix} P_{1m} & P_{2m} \\ * & P_{3m} \end{bmatrix}, \ R_s = \begin{bmatrix} R_{1s} & X_s \\ R_{2s} & X_s \end{bmatrix}, \\ \bar{P}_m &= \begin{bmatrix} \bar{P}_{1m} & \bar{P}_{2m} \\ * & \bar{P}_{3m} \end{bmatrix} = \sum_{n=1}^{M} \vartheta_{mn} \begin{bmatrix} P_{1n} & P_{2n} \\ * & P_{3n} \end{bmatrix}, \\ \left[\begin{array}{c|c} \mathcal{A}_{si} & \mathcal{B}_{si} \\ \hline \mathcal{E}_{si} & \mathcal{F}_{si} \end{array}\right] &= \left[\begin{array}{c|c} X_s A_{fsi} & X_s B_{fsi} \\ \hline E_{fsi} & F_{fsi} \end{array}\right], \\ G_s &= \begin{bmatrix} G_{1s} & X_s \\ G_{2s} & X_s \end{bmatrix}. \end{aligned} \tag{11.29}$$

By computation, it is easy to find that LMIs (11.26)–(11.27) are equivalent to LMIs (11.9)–(11.10). If there is an solution to LMIs (11.25)–(11.27), the filter gains can be obtained by (11.28). The proof is completed.

In the following theorem, we show how to eliminate these unimportant matrices including G_{1s}, G_{2s}, R_{1s} and R_{2s}, and then another filtering design algorithm with less conservatism is proposed.

Theorem 11.3 *If there exist matrices $P_m = \begin{bmatrix} P_{1m} & P_{2m} \\ * & P_{3m} \end{bmatrix} > 0$, X_s, H and symmetric matrix W_{msi} for any $i, \ j \in \mathcal{I}$, $m \in \mathcal{M}$ and $s \in \mathcal{S}$ satisfying*

$$\sum_{s=1}^{S} \pi_{ms} W_{msi} < \mathcal{V} - \alpha I, \tag{11.30}$$

$$(\Omega_{mi}^1)^T \Xi_{msii} \Omega_{mi}^1 < 0, \tag{11.31}$$

$$\Theta^T \Xi_{msii} \Theta < 0, \tag{11.32}$$

$$(\Omega_{mij}^2)^T (\Xi_{msij} + \Xi_{msji}) \Omega_{mij}^2 < 0, \ (i < j), \tag{11.33}$$

$$\Theta^T (\Xi_{msij} + \Xi_{msji}) \Theta < 0, \ (i < j), \tag{11.34}$$

where

$$\Xi_{msij} = \begin{bmatrix} 0 & -X_s & \Xi^{13}_{msij} & \Xi^{14}_{msij} & \Xi^{15}_{msij} & 0 \\ * & \Xi^{22}_{s} & \Xi^{23}_{msij} & \Xi^{24}_{msij} & \Xi^{25}_{msij} & 0 \\ * & * & \Xi^{33}_{msij} & \Xi^{34}_{msij} & \Xi^{35}_{msij} & \Xi^{36}_{msij} \\ * & * & * & \Xi^{44}_{msij} & \Xi^{45}_{msij} & \Xi^{46}_{msij} \\ * & * & * & * & -W_{msi} & \Xi^{56}_{msij} \\ * & * & * & * & * & \Xi^{66}_{msij} \end{bmatrix},$$

$$\begin{aligned}
&\Xi^{22}_{s} = -\mathbf{Her}(X_s),\ \Xi^{23}_{msij} = P^T_{2m} + \mathcal{B}_{fsj}C_{mi} - X^T_s,\\
&\Xi^{13}_{msij} = P_{1m} + \mathcal{B}_{fsj}C_{mi},\ \Xi^{14}_{msij} = P_{2m} + \mathcal{A}_{fsj},\\
&\Xi^{33}_{msij} = \bar{P}_{1m} + \mathbf{Her}(\mathcal{B}_{fsj}C_{mi}),\ \Xi^{36}_{msij} = \Gamma^{36}_{msij},\\
&\Xi^{24}_{msij} = P_{3m} + \mathcal{A}_{fsj} - X^T_s,\ \Xi^{46}_{msij} = \Gamma^{46},\\
&\Xi^{34}_{msij} = \bar{P}_{2m} + \mathcal{A}_{fsj} + (\mathcal{B}_{fsj}C_{mi})^T,\ \Xi^{56}_{msij} = \Gamma^{56}_{msij},\\
&\Xi^{44}_{msij} = \bar{P}_{3m} + \mathbf{Her}(\mathcal{A}_{fsj}),\ \Xi^{66}_{msij} = \Gamma^{66}_{msij},\\
&\Xi^{15}_{msij} = \Xi^{25}_{msij} = \Xi^{35}_{msij} = \Xi^{45}_{msij} = \mathcal{B}_{fsj}D_{mi},\\
&\Omega^{1}_{mi} = \begin{bmatrix} \Omega^{11}_{mi} \\ \Omega^{12} \end{bmatrix},\ \Omega^{2}_{mij} = \begin{bmatrix} \frac{\Omega^{11}_{mi}+\Omega^{11}_{mj}}{2} \\ \Omega^{12} \end{bmatrix},\\
&\Theta = \begin{bmatrix} 0\ 0\ 0\ 0\ \Theta^1 \end{bmatrix}^T,\ \Theta^1 = diag\{I,\ I\},\\
&\Omega^{11}_{mi} = \begin{bmatrix} 0\ A_{mi}\ 0\ B_{mi}\ 0 \end{bmatrix},\ \Omega^{12} = diag\{I,\ I,\ I,\ I,\ I\},
\end{aligned}$$

the filtering error system in (11.6) is stochastically stable with strict $(\mathcal{U},\ \mathcal{G},\ \mathcal{V})$-$\alpha$ dissipativity. Furthermore, we can derive the asynchronous dissipative fuzzy filtering parameters from (11.28).

Proof Note that (11.26) can be rewritten as

$$\Gamma_{msii} = \Xi_{msii} + \mathbf{Her}(\Sigma \Delta_s \Pi_{mi}) < 0, \tag{11.35}$$

And

$$\begin{aligned}\Gamma_{msij} + \Gamma_{msji} = \Xi_{msij} + \Xi_{msji} +& \\ \mathbf{Her}(\Sigma \Delta_s (\Pi_{mi} + \Pi_{mj})) < 0,&\end{aligned} \tag{11.36}$$

where

$$\Sigma = \begin{bmatrix} I\ 0\ 0\ 0\ 0\ 0 \\ 0\ I\ 0\ 0\ 0\ 0 \\ 0\ 0\ I\ 0\ 0\ 0 \\ 0\ 0\ 0\ I\ 0\ 0 \end{bmatrix}^T,\ \Delta_s = \begin{bmatrix} R_{1s} \\ R_{2s} \\ G_{1s} \\ G_{2s} \end{bmatrix},$$

$$\Pi_{mi} = \begin{bmatrix} -I\ 0\ A_{mi}\ 0\ B_{mi}\ 0 \end{bmatrix}.$$

We can clearly observe that Ω_{mi}^1, Ω_{mij}^2 and Θ are the right null space of Π_{mi}, $\Pi_{mi} + \Pi_{mj}$ and Σ^T, respectively. From Finsler's Lemma, the feasibility of LMI (11.26) is equivalent to that of LMIs (11.31)–(11.32). We can also derive the equivalent relationship between LMI (11.27) and LMIs (11.33)–(11.34). The proof is completed.

Remark 11.4 Both Theorems 11.2 and 11.3 provide dissipative asynchronous filtering design approaches for continuous-time T–S fuzzy MJSs. By constructing appropriate slack matrices and the Lyapunov function matrix, the dissipative asynchronous filtering issue is converted into a convex optimization problem. We can achieve the optimal dissipative performance α^* by minimizing $-\alpha$ subject to LMIs in Theorems 11.2 and 11.3, respectively. The corresponding filter parameters can be obtained through (11.28). By using Finsler's lemma, unnecessary matrices R_{1s}, R_{2s}, G_{1s} and G_{2s} are removed, and the number of decision variables drops by $4n_x^2 S$, which means that Theorem 11.3 has fewer constraints and less conservatism. However, the number of LMIs increases by $0.5MSr(r+1)$, and the amount of computation is added to some extent.

11.4 Conclusion

The dissipative asynchronous filtering issue has been investigated for continuous-time T–S fuzzy MJSs, whose asynchronous phenomenon has been described by the HMM theory. Two LMI-based methods have been developed for designing an asynchronous fuzzy filter that guarantee the stochastic stability of the filtering error system with a given dissipative performance criterion.

Chapter 12
Networked Fault Detection for Fuzzy MJSs

12.1 Introduction

In realistic control systems, faults are significant obstacles to obtain a better performance, higher reliability and safety [1–3]. They would likely appear due to unexpected variations in signals and components, sudden changes of working conditions, environmental noises, parameter shifting, etc. It is important to detect them timely and accurately for avoiding the degradation of performance and instability. The model-based fault detection (FD) method has been proposed, whose primary goals are to construct an appropriate filter/observer, generate the residual signal and evaluate the produced residual signal. A fault alarm will be sent out once the residual evaluation value is larger than the predefined threshold. Recently, plenty of dynamical systems have adopted this approach to detect faults and guarantee the system's normal operation. Through the delta operator method, the work in [4] has analyzed the FD issue for T–S fuzzy systems with time-varying delays. The FD filtering problem has been addressed in [5] for nonlinear switched stochastic systems. For T–S fuzzy systems with unknown bounded noises and sensor faults, the work in [6] has considered the FD observer design as a multi-objective H_-/H_∞ performance index. An event-triggered method has been used in [7] to investigate the FD problem for nonlinear networked systems.

In this chapter, we investigate the dissipative asynchronous FD problem for T–S fuzzy MJSs with network data losses. We assume that there are imperfect communication links between the plant and the designed FD filter, described by Bernoulli process. Besides, the HMM theory is applied to describe the non-synchronization between two. Then based on the fuzzy inference, a fuzzy FD filter is devised to produce a residual signal and the FD issue is transformed as a filtering problem. Via the mode-dependent and fuzzy-basis-dependent Lyapunov function technique, a sufficient condition is developed to ensure the stochastic stability and the strict dissipativity of the FD system. Two approaches are proposed to obtain FD filter gains, which can be solved by using Matlab Toolbox.

S. Dong et al., *Control and Filtering of Fuzzy Systems with Switched Parameters*, Studies in Systems, Decision and Control 268,
https://doi.org/10.1007/978-3-030-35566-1_12

12.2 Preliminary Analysis

Consider the following discrete-time T–S fuzzy systems with Markov jump:
Plant rule i: IF μ_{1k} is ϕ_{i1}, μ_{2k} is ϕ_{i2}, ..., and μ_{pk} is ϕ_{ip}, THEN

$$\begin{cases} x_{k+1} = A_{\delta_k i} x_k + B_{\delta_k i} u_k + C_{\delta_k i} w_k + D_{\delta_k i} f_k, \\ y_k = E_{\delta_k i} x_k + F_{\delta_k i} u_k + G_{\delta_k i} w_k + H_{\delta_k i} f_k, \end{cases} \tag{12.1}$$

where $x_k \in R^{n_x}$ is the state variable; $u_k \in R^{n_u}$ is the given input; $w_k \in R^{n_w}$ is the external disturbance which belongs to $l_2[0, +\infty)$; $f_k \in R^{n_f}$ is the known fault signal; and $y_k \in R^{n_y}$ is the measured output. System matrices are known with appropriate dimensions. The variable i ($i \in \mathcal{I} = \{1, 2, \ldots, r\}$) means the ith fuzzy rule and r is the total sum of rules. ϕ_{ij} ($j \in \mathcal{J} = \{1, 2, \ldots, p\}$) is the fuzzy set and μ_{jk} is the premise variable. δ_k is adopted to represent a discrete-time Markov jump process, taking values in $\mathcal{V} = \{1, 2, \ldots, V\}$. δ_k is subject to the transition probability matrix $\Upsilon = [\lambda_{ab}]$ with

$$\Pr\{\delta_{k+1} = b | \delta_k = a\} = \lambda_{ab}, \; a, \; b \in \mathcal{V}, \tag{12.2}$$

where $\lambda_{ab} \geq 0$ and $\sum_{b=1}^{V} \lambda_{ab} = 1$.

With $\delta_k = a$, the overall fuzzy systems can be inferred as

$$\begin{cases} x_{k+1} = A_{ah} x_k + B_{ah} u_k + C_{ah} w_k + D_{ah} f_k, \\ y_k = E_{ah} x_k + F_{ah} u_k + G_{ah} w_k + H_{ah} f_k, \end{cases} \tag{12.3}$$

where

$$A_{ah} = \sum_{i=1}^{r} h_i(\mu_k) A_{ai}, \; B_{ah} = \sum_{i=1}^{r} h_i(\mu_k) B_{ai},$$

$$C_{ah} = \sum_{i=1}^{r} h_i(\mu_k) C_{ai}, \; D_{ah} = \sum_{i=1}^{r} h_i(\mu_k) D_{ai},$$

$$E_{ah} = \sum_{i=1}^{r} h_i(\mu_k) E_{ai}, \; F_{ah} = \sum_{i=1}^{r} h_i(\mu_k) F_{ai},$$

$$G_{ah} = \sum_{i=1}^{r} h_i(\mu_k) G_{ai}, \; H_{ah} = \sum_{i=1}^{r} h_i(\mu_k) H_{ai},$$

$$\mu_k = \left[\mu_{1k}, \mu_{2k}, \ldots, \mu_{pk}\right], \; h_i(\mu_k) = \frac{\prod_{j=1}^{p} \phi_{ij}(\mu_{jk})}{\sum_{i=1}^{r} \prod_{j=1}^{p} \phi_{ij}(\mu_{jk})}.$$

$\phi_{ij}(\mu_{jk})$ denotes the grade of membership of μ_{jk} in ϕ_{ij} and we assume that $\prod_{j=1}^{p} \phi_{ij}(\mu_{jk}) \geq 0$. Accordingly, it is observed easily that $h_i(\mu_k) \geq 0$ and $\sum_{i=1}^{r} h_i(\mu_k) = 1$. In the following, we represent $h_i(\mu_k)$ as h_i for convenient expression.

Owing to communication link constraints between the plant and an FD filter to be designed, data losses may happen stochastically. We adopt the following model to represent the random data dropout phenomenon:

$$y_{fk} = \beta_k y_k, \tag{12.4}$$

where β_k represents Bernoulli process. $\beta_k = 1$ means that data are transmitted successfully. Otherwise, there will be transmission failure if $\beta_k = 0$. We suppose that β_k is subject to

$$\Pr\{\beta_k = 1\} = \beta, \ \Pr\{\beta_k = 0\} = 1 - \beta. \tag{12.5}$$

It easily follows that $E\{\beta_k\} = \beta$. Define $\bar{\beta}_k = \beta_k - \beta$, and we have

$$E\{\bar{\beta}_k\} = 0, \ E\{\bar{\beta}_k^2\} = \beta(1-\beta) = \bar{\beta}^2 \tag{12.6}$$

with $\bar{\beta} = \sqrt{\beta(1-\beta)}$ and $\beta \in (0, 1]$.

In this chapter, we are interested in designing an FD filter to generate the residual signal that is sensitive to faults. Besides, it is supposed that premise variables of the designed filter are the same as those of the plant. Applying the PDC approach, we devise the following fuzzy filter:

Filter rule i: IF μ_{1k} is ϕ_{i1}, μ_{2k} is ϕ_{i2}, ..., and μ_{pk} is ϕ_{ip}, THEN

$$\begin{cases} \hat{x}_{k+1} = \hat{A}_{\rho_k i}\hat{x}_k + \hat{B}_{\rho_k i} y_{fk}, \\ r_k = \hat{E}_{\rho_k i} x_k + \hat{F}_{\rho_k i} y_{fk}, \end{cases} \tag{12.7}$$

where $\hat{x}_k \in R^{n_{\hat{x}}}$ is the state variable; $r_k \in R^{n_r}$ is the residual signal; Filter matrices $(\hat{A}_{\rho_k i}, \ \hat{B}_{\rho_k i}, \ \hat{E}_{\rho_k i}, \ \hat{F}_{\rho_k i})$ are to be solved. The variable ρ_k is introduced to observe the plant mode, which is subject to the conditional transition probability matrix $\Phi = [\varphi_{as}]$ ($\rho_k \in \mathcal{L} = \{1, 2, \ldots, L\}$) with

$$\Pr\{\rho_k = s | \delta_k = a\} = \varphi_{as}, \tag{12.8}$$

where $\varphi_{as} > 0$ and $\sum_{s=1}^{L} \varphi_{as} = 1$. Therefore, when $\rho_k = s$, we can represent the filter as

$$\begin{cases} \hat{x}_{k+1} = \hat{A}_{sh}\hat{x}_k + \hat{B}_{sh} y_{fk}, \\ r_k = \hat{E}_{sh}\hat{x}_k + \hat{F}_{sh} y_{fk}, \end{cases} \tag{12.9}$$

where

$$\hat{A}_{sh} = \sum_{i=1}^{r} h_i \hat{A}_{si}, \ \hat{B}_{sh} = \sum_{i=1}^{r} h_i \hat{B}_{si},$$

$$\hat{E}_{sh} = \sum_{i=1}^{r} h_i \hat{E}_{si}, \ \hat{F}_{sh} = \sum_{i=1}^{r} h_i \hat{F}_{si}.$$

For FD systems, we introduce a reference model to obtain a better performance, that is, $\tilde{f}(z) = W(z)f(z)$, where $W(z)$ is a known stable weighting matrix. The minimal state-space realization is expressed as

$$\begin{cases} \tilde{x}_{k+1} = A_W\tilde{x}_k + B_W f_k, \\ \tilde{f}_k = E_W\tilde{x}_k + F_W f_k, \end{cases} \tag{12.10}$$

where $\tilde{x}_k \in R^{n_{\tilde{x}}}$ is the state of weighted fault; $\tilde{f}_k \in R^{n_{\tilde{f}}}$ is the weighted fault; And (A_W, B_W, E_W, F_W) are known constant matrices.

Defining the residual error as $e_k = r_k - \tilde{f}_k$, combining systems (12.3), (12.9) and (12.10), we obtain the following FD system:

$$\begin{cases} \zeta_{k+1} = (\breve{A}_{1ash} + \bar{\beta}_k\breve{A}_{2ash})\zeta_k + (\breve{B}_{1ash} + \bar{\beta}_k\breve{B}_{2ash})\upsilon_k, \\ e_k = (\breve{E}_{1ash} + \bar{\beta}_k\breve{E}_{2ash})\zeta_k + (\breve{F}_{1ash} + \bar{\beta}_k\breve{F}_{2ash})\upsilon_k, \end{cases} \tag{12.11}$$

where

$$\breve{A}_{1ash} = \sum_{i=1}^{r}\sum_{j=1}^{r} h_i h_j \breve{A}_{1asij}, \ \breve{A}_{1asij} = \begin{bmatrix} \bar{A}_{ai} & 0 \\ \beta\hat{B}_{sj}\bar{E}_{ai} & \hat{A}_{sj} \end{bmatrix},$$

$$\breve{A}_{2ash} = \sum_{i=1}^{r}\sum_{j=1}^{r} h_i h_j \breve{A}_{2asij}, \ \breve{A}_{2asij} = \begin{bmatrix} 0 & 0 \\ \hat{B}_{sj}\bar{E}_{ai} & 0 \end{bmatrix},$$

$$\breve{B}_{1ash} = \sum_{i=1}^{r}\sum_{j=1}^{r} h_i h_j \breve{B}_{1asij}, \ \breve{B}_{1asij} = \begin{bmatrix} \bar{B}_{ai} \\ \beta\hat{B}_{sj}\bar{F}_{ai} \end{bmatrix},$$

$$\breve{B}_{2ash} = \sum_{i=1}^{r}\sum_{j=1}^{r} h_i h_j \breve{B}_{2asij}, \ \breve{B}_{2asij} = \begin{bmatrix} 0 \\ \hat{B}_{sj}\bar{F}_{ai} \end{bmatrix},$$

$$\breve{E}_{2ash} = \sum_{i=1}^{r}\sum_{j=1}^{r} h_i h_j \breve{E}_{2asij}, \ \breve{E}_{2asij} = \begin{bmatrix} \hat{F}_{sj}\bar{E}_{ai} & 0 \end{bmatrix},$$

$$\breve{F}_{1ash} = \sum_{i=1}^{r}\sum_{j=1}^{r} h_i h_j \breve{F}_{1asij}, \ \breve{F}_{1asij} = \beta\hat{F}_{sj}\bar{F}_{ai} - \bar{F}_W,$$

$$\breve{F}_{2ash} = \sum_{i=1}^{r}\sum_{j=1}^{r} h_i h_j \breve{F}_{2asij}, \ \breve{F}_{2asij} = \hat{F}_{sj}\bar{F}_{ai},$$

$$\breve{E}_{1ash} = \sum_{i=1}^{r}\sum_{j=1}^{r} h_i h_j \breve{E}_{1asij}, \ \bar{F}_W = \begin{bmatrix} 0 & 0 & F_W \end{bmatrix},$$

$$\breve{E}_{1asij} = \begin{bmatrix} \beta\hat{F}_{sj}\bar{E}_{ai} - \bar{E}_W & \hat{E}_{sj} \end{bmatrix}, \ \bar{E}_W = \begin{bmatrix} E_W & 0 \end{bmatrix},$$

$$\bar{A}_{ai} = \begin{bmatrix} A_W & 0 \\ 0 & A_{ai} \end{bmatrix}, \ \bar{B}_{ai} = \begin{bmatrix} 0 & 0 & B_W \\ B_{ai} & C_{ai} & D_{ai} \end{bmatrix},$$

$$\bar{E}_{ai} = \begin{bmatrix} 0 & E_{ai} \end{bmatrix}, \; \bar{F}_{ai} = \begin{bmatrix} F_{ai} & G_{ai} & H_{ai} \end{bmatrix},$$
$$\upsilon_k = \begin{bmatrix} u_k^T & w_k^T & f_k^T \end{bmatrix}^T, \; \zeta_k = \begin{bmatrix} \tilde{x}_k^T & x_k^T & \hat{x}_k^T \end{bmatrix}^T.$$

The FD problem of this chapter can be represented as the following two steps:
(1) Produce the residual signal: for T–S fuzzy MJSs (12.3), design a suitable dissipative fuzzy FD filter (12.7) to generate the residual signal. Furthermore, the designed filter can guarantee that the FD system (12.11) is stochastically stable with strict dissipative performance;
(2) Construct an FD measure: calculate the residual evaluation value and the threshold via a chosen evaluation function. If the threshold is smaller than the residual evaluation value, a fault alarm is sent out. In this chapter, we select the evaluation function $\mathbf{J}(r)$ and the threshold $\mathbf{J}_{th}(r)$ as

$$\mathbf{J}(r) = \sqrt{\sum_{i=k_0}^{k_0+k} r_i^T r_i}, \; \mathbf{J}_{th}(r) = \sup_{w\neq 0, u\neq 0, f=0} \mathbf{J}(r), \tag{12.12}$$

where k_0 is the initial evaluation time. Via the following test, the occurrence of fault can be detected:

$$\begin{aligned} &\mathbf{J}(r) \leq \mathbf{J}_{th}(r) \Longrightarrow \text{no faults}, \\ &\mathbf{J}(r) > \mathbf{J}_{th}(r) \Longrightarrow \text{faults} \Longrightarrow \text{alarm}. \end{aligned} \tag{12.13}$$

12.3 Main Results

In this section, a sufficient condition is developed to ensure the stochastic stability and the strict dissipativity for the FD system (12.11) on the supposition that the fuzzy FD filter matrices in (12.7) are known. Then, two asynchronous FD design approaches are given.

Theorem 12.1 *The FD system (12.11) is stochastically stable and strictly ($\mathcal{Q}$, $\mathcal{S}$, $\mathcal{R}$)-α-dissipative if there exist n-dimensional matrices $P_{ai} > 0$ and $M_{asi} > 0$ ($n = n_{\tilde{x}} + n_x + n_{\hat{x}}$) for any $a \in \mathcal{V}$, $s \in \mathcal{L}$ and $i, j, t \in \mathcal{I}$ subject to*

$$\sum_{s=1}^{L} \varphi_{as} M_{asi} < P_{ai}, \tag{12.14}$$

$$\Gamma_{asiit} < 0, \tag{12.15}$$

$$\Gamma_{asijt} + \Gamma_{asjit} < 0, \; i < j, \tag{12.16}$$

where

$$\Gamma_{asijt} = \begin{bmatrix} -\Gamma_{11} & 0 & \Gamma_{11}\Gamma_{13} & \Gamma_{11}\Gamma_{14} \\ * & -\Gamma_{22} & \Gamma_{23}^{1}\Gamma_{23} & \Gamma_{23}^{1}\Gamma_{24} \\ * & * & -M_{asi} & -\breve{E}_{1asij}^{T}\mathcal{S} \\ * & * & * & \Gamma_{44} \end{bmatrix},$$

$$\Gamma_{11} = diag\{\bar{P}_{at}, \bar{P}_{at}\},\ \Gamma_{23}^{1} = diag\{\mathcal{Q}_{-}, \mathcal{Q}_{-}\},$$

$$\Gamma_{13} = \begin{bmatrix} \breve{A}_{1asij} \\ \bar{\beta}\breve{A}_{2asij} \end{bmatrix},\ \Gamma_{14} = \begin{bmatrix} \breve{B}_{1asij} \\ \bar{\beta}\breve{B}_{2asij} \end{bmatrix},$$

$$\Gamma_{23} = \begin{bmatrix} \breve{E}_{1asij} \\ \bar{\beta}\breve{E}_{2asij} \end{bmatrix},\ \Gamma_{24} = \begin{bmatrix} \breve{F}_{1asij} \\ \bar{\beta}\breve{F}_{2asij} \end{bmatrix},$$

$$\Gamma_{44} = -\breve{F}_{1asij}^{T}\mathcal{S} - \mathcal{S}^{T}\breve{F}_{1asij} - \mathcal{R} + \alpha I,$$

$$\bar{P}_{at} = \sum_{b=1}^{V}\lambda_{ab}P_{bt},\ \Gamma_{22} = diag\{I, I\}.$$

Proof Define

$$\begin{aligned} M_{ash} &= \sum_{i=1}^{r} h_i M_{asi},\ P_{ah} = \sum_{i=1}^{r} h_i P_{ai}, \\ P_{bh^+} &= \sum_{t=1}^{r} h_t^+ P_{bt},\ \bar{P}_{ah^+} = \sum_{b=1}^{V}\lambda_{ab}P_{bh^+},\ h^+ = h(\mu_{k+1}). \end{aligned} \tag{12.17}$$

According to the fuzzy principle, (12.11) and (12.14)–(12.16), we have

$$\sum_{s=1}^{L}\varphi_{as}M_{ash} < P_{ah}, \tag{12.18}$$

and

$$\begin{aligned} \Gamma_{ash} &= \sum_{t=1}^{r}\sum_{i=1}^{r}\sum_{j=1}^{r} h_t^+ h_i h_j \Gamma_{asijt} \\ &= \sum_{t=1}^{r} h_t^+ \left(\sum_{i=1}^{r} h_i^2 \Gamma_{asiit} + \sum_{i=1}^{r-1}\sum_{j=i+1}^{r} h_i h_j (\Gamma_{asijt} + \Gamma_{asjit}) \right) < 0, \end{aligned} \tag{12.19}$$

where

$$\Gamma_{ash} = \begin{bmatrix} -\Gamma_{11h^+} & 0 & \Gamma_{11h^+}\Gamma_{13h} & \Gamma_{11h^+}\Gamma_{14h} \\ * & -\Gamma_{22} & \Gamma_{23}^{1}\Gamma_{23h} & \Gamma_{23}^{1}\Gamma_{24h} \\ * & * & -M_{ash} & -\breve{E}_{1ash}^{T}\mathcal{S} \\ * & * & * & \Gamma_{44h} \end{bmatrix},$$

$$\begin{aligned}
\Gamma_{11h^+} &= \text{diag}\{\bar{P}_{ah^+}, \bar{P}_{ah^+}\},\\
\Gamma_{13h} &= \begin{bmatrix} \breve{A}_{1ash} \\ \bar{\beta}\breve{A}_{2ash} \end{bmatrix},\ \Gamma_{14h} = \begin{bmatrix} \breve{B}_{1ash} \\ \bar{\beta}\breve{B}_{2ash} \end{bmatrix},\\
\Gamma_{23h} &= \begin{bmatrix} \breve{E}_{1ash} \\ \bar{\beta}\breve{E}_{2ash} \end{bmatrix},\ \Gamma_{24h} = \begin{bmatrix} \breve{F}_{1ash} \\ \bar{\beta}\breve{F}_{2ash} \end{bmatrix},\\
\Gamma_{44h} &= -\breve{F}_{1ash}^T\mathcal{S} - \mathcal{S}^T\breve{F}_{1ash} - \mathcal{R} + \alpha I.
\end{aligned}$$

Applying Schur Complement to (12.19), and considering $-\mathcal{Q} = \mathcal{Q}_-^T\mathcal{Q}_-$, we have

$$\begin{cases} \Gamma_{13h}^T\Gamma_{11h^+}\Gamma_{13h} - M_{ash} < 0, \\ \Gamma_{134h}^T\Gamma_{11h^+}\Gamma_{134h} - \Gamma_{234h}^T\Gamma_{23}^{1'}\Gamma_{234h} + \Gamma_{44h}' < 0, \end{cases} \tag{12.20}$$

where

$$\begin{aligned}
\Gamma_{134h} &= \begin{bmatrix}\Gamma_{13h} & \Gamma_{14h}\end{bmatrix},\ \Gamma_{234h} = \begin{bmatrix}\Gamma_{23h} & \Gamma_{24h}\end{bmatrix},\\
\Gamma_{23}^{1'} &= \text{diag}\{\mathcal{Q}, \mathcal{Q}\},\ \Gamma_{44h}' = \begin{bmatrix} -M_{ash} & -\breve{E}_{1ash}^T\mathcal{S} \\ * & \Gamma_{44h} \end{bmatrix}.
\end{aligned}$$

Furthermore, from (12.18), it follows that

$$\Delta_i < 0,\ i = 1, 2, \tag{12.21}$$

where

$$\begin{aligned}
\Delta_1 &= \sum_{s=1}^{L}\varphi_{as}\left(\Gamma_{13h}^T\Gamma_{11h^+}\Gamma_{13h} - P_{ah}\right),\\
\Delta_2 &= \sum_{s=1}^{L}\varphi_{as}\left(\Gamma_{134h}^T\Gamma_{11h^+}\Gamma_{134h} - \Gamma_{234h}^T\Gamma_{23}^{1'}\Gamma_{234h} + \Gamma_{44h}''\right),\\
\Gamma_{44h}'' &= \begin{bmatrix} -P_{ah} & -\breve{E}_{1ash}^T\mathcal{S} \\ * & \Gamma_{44h} \end{bmatrix}.
\end{aligned}$$

We choose the Lyapunov function as

$$V_k = \zeta_k^T P_{\delta_k h}\zeta_k, \tag{12.22}$$

where $P_{\delta_k h} = \sum_{i=1}^{r} h_i P_{\delta_k i}$, and $P_{\delta_k i}$ is mode-dependent and fuzzy-basis-dependent. It is assumed that at time k, $\delta_k = a$. At the next time $k+1$, $\delta_{k+1} = b$ and the Lyapunov function matrix turns to be P_{bh^+}.

Along the trajectory of system (12.11) with $\upsilon_k = 0$, we have

$$\begin{aligned}
E\{\Delta V_k\} &= E\{\zeta_{k+1}^T P_{bh^+}\zeta_{k+1}\} - E\{\zeta_k^T P_{ah}\zeta_k\}\\
&= \zeta_k^T\Delta_1\zeta_k < 0.
\end{aligned} \tag{12.23}$$

It is clearly concluded that system (12.11) is stochastically stable.

Now, we are going to establish the strict dissipativity performance for system (12.11).

The energy supply function for system (12.11) is denoted as

$$F_{dis}(e_k, \upsilon_k) = e_k^T \mathcal{Q} e_k + 2e_k^T \mathcal{S} \upsilon_k + \upsilon_k^T \mathcal{R} \upsilon_k, \tag{12.24}$$

where matrices $\mathcal{Q}$, $\mathcal{S}$ and $\mathcal{R}$ are known with $\mathcal{R}^T = \mathcal{R}$. And $\mathcal{Q}$ is a negative semi-definite matrix, which implies that $-\mathcal{Q} = \mathcal{Q}_-^T \mathcal{Q}_-$.

From (12.21), it follows that

$$E\{\Delta V_k - F_{dis}(e_k, \upsilon_k) + \alpha \upsilon_k^T \upsilon_k\} = \begin{bmatrix} \zeta_k \\ \upsilon_k \end{bmatrix}^T \Delta_2 \begin{bmatrix} \zeta_k \\ \upsilon_k \end{bmatrix} < 0. \tag{12.25}$$

Summing up the above inequality from $k = 0$ to K yields that

$$V_{K+1} - V_0 - \sum_{k=0}^{K} E\{F_{dis}(e_k, \upsilon_k)\} + \alpha \sum_{k=0}^{K} \upsilon_k^T \upsilon_k < 0. \tag{12.26}$$

With the zero initial condition, namely, $V_0 = 0$, it follows that

$$\sum_{k=0}^{K} E\{F_{dis}(e_k, \upsilon_k)\} > \alpha \sum_{k=0}^{K} \upsilon_k^T \upsilon_k + V_{K+1} > \alpha \sum_{k=0}^{K} \upsilon_k^T \upsilon_k. \tag{12.27}$$

We can obtain that system (12.11) is strictly dissipative, which completes the proof.

Based on Theorem 12.1, we focus on designing an $(n_{\tilde{x}} + n_x)$-dimensional asynchronous FD filter for T–S fuzzy MJSs with data losses in the following.

Theorem 12.2 *If there exist n-dimensional matrices* $P_{ai} = \begin{bmatrix} P_{1ai} & P_{2ai} \\ * & P_{3ai} \end{bmatrix} > 0$, $M_{asi} = \begin{bmatrix} M_{1asi} & M_{2asi} \\ * & M_{3asi} \end{bmatrix}$ $(n = 2(n_{\tilde{x}} + n_x))$, *$(n_{\tilde{x}} + n_x)$-dimensional matrices* N_{1s}, N_{2s}, X_s, $\hat{\mathscr{A}}_{sj}$, $\hat{\mathscr{B}}_{sj}$, $\hat{\mathscr{E}}_{sj}$, *and* $\hat{\mathscr{F}}_{sj}$ *for any* $a \in \mathcal{V}$, $s \in \mathcal{L}$ *and* $i, j, t \in \mathcal{I}$ *subject to*

$$\sum_{s=1}^{L} \varphi_{as} M_{asi} < P_{ai}, \tag{12.28}$$

$$\Lambda_{asiit} < 0, \tag{12.29}$$

$$\Lambda_{asijt} + \Lambda_{asjit} < 0, \; i < j, \tag{12.30}$$

where

$$\Lambda_{asijt} = \begin{bmatrix} -\Lambda_{11} & 0 & 0 & \Lambda_{14} & \Lambda_{15} \\ * & -\Lambda_{11} & 0 & \Lambda_{24} & \Lambda_{25} \\ * & * & -\Lambda_{33} & \Lambda_{34} & \Lambda_{35} \\ * & * & * & -M_{asi} & \Lambda_{45} \\ * & * & * & * & \Lambda_{55} \end{bmatrix},$$

$$\Lambda_{11} = \begin{bmatrix} \bar{P}_{1at} - N_{1s} - N_{1s}^T & \bar{P}_{2at} - X_s - N_{2s}^T \\ * & \bar{P}_{3at} - X_s - X_s^T \end{bmatrix},$$

$$\Lambda_{14} = \begin{bmatrix} N_{1s}\bar{A}_{ai} + \beta\hat{\mathscr{B}}_{sj}\bar{E}_{ai} & \hat{\mathscr{A}}_{sj} \\ N_{2s}\bar{A}_{ai} + \beta\hat{\mathscr{B}}_{sj}\bar{E}_{ai} & \hat{\mathscr{A}}_{sj} \end{bmatrix},$$

$$\Lambda_{15} = \begin{bmatrix} N_{1s}\bar{B}_{ai} + \beta\hat{\mathscr{B}}_{sj}\bar{F}_{ai} \\ N_{2s}\bar{B}_{ai} + \beta\hat{\mathscr{B}}_{sj}\bar{F}_{ai} \end{bmatrix}, \ \Lambda_{33} = diag\{I, I\},$$

$$\Lambda_{24} = \begin{bmatrix} \bar{\beta}\hat{\mathscr{B}}_{sj}\bar{E}_{ai} & 0 \\ \bar{\beta}\hat{\mathscr{B}}_{sj}\bar{E}_{ai} & 0 \end{bmatrix}, \ \Lambda_{25} = \begin{bmatrix} \bar{\beta}\hat{\mathscr{B}}_{sj}\bar{F}_{ai} \\ \bar{\beta}\hat{\mathscr{B}}_{sj}\bar{F}_{ai} \end{bmatrix},$$

$$\Lambda_{34} = \begin{bmatrix} \mathcal{Q}_{-}(\beta\hat{\mathscr{F}}_{sj}\bar{E}_{ai} - \bar{E}_W) & \mathcal{Q}_{-}\hat{\mathscr{E}}_{sj} \\ \bar{\beta}\mathcal{Q}_{-}\hat{\mathscr{F}}_{sj}\bar{E}_{ai} & 0 \end{bmatrix},$$

$$\Lambda_{35} = \begin{bmatrix} \mathcal{Q}_{-}(\beta\hat{\mathscr{F}}_{sj}\bar{F}_{ai} - \bar{F}_W) \\ \bar{\beta}\mathcal{Q}_{-}\hat{\mathscr{F}}_{sj}\bar{F}_{ai} \end{bmatrix},$$

$$\Lambda_{45} = \begin{bmatrix} -(\beta\hat{\mathscr{F}}_{sj}\bar{E}_{ai} - \bar{E}_W)^T\mathcal{S} \\ -\hat{\mathscr{E}}_{sj}^T\mathcal{S} \end{bmatrix},$$

$$\Lambda_{55} = -(\beta\hat{\mathscr{F}}_{sj}\bar{F}_{ai} - \bar{F}_W)^T\mathcal{S} - \mathcal{S}^T(\beta\hat{\mathscr{F}}_{sj}\bar{F}_{ai} - \bar{F}_W) - \mathcal{R} + \alpha I,$$

there exists an $(n_{\tilde{x}} + n_x)$*-dimensional dissipative FD filter in the form of (12.7) with* $(n_{\hat{x}} = n_{\tilde{x}} + n_x)$*, which can guarantee that the FD system (12.11) is stochastically stable with strict dissipativity. Moreover, we can obtain the filter gains as follows:*

$$\begin{cases} \hat{A}_{sj} = X_s^{-1}\hat{\mathscr{A}}_{sj}, \ \hat{E}_{sj} = \hat{\mathscr{E}}_{sj}, \\ \hat{B}_{sj} = X_s^{-1}\hat{\mathscr{B}}_{sj}, \ \hat{F}_{sj} = \hat{\mathscr{F}}_{sj}. \end{cases} \tag{12.31}$$

Proof Pre-multiplying $\text{diag}\{\Gamma_{11}^1\Gamma_{11}^{-1}, I, I, I\}$ and post-multiplying its transpose to (12.15) with $\Gamma_{11}^1 = \text{diag}\{N_s, N_s\}$, we have

$$\Gamma'_{asiit} = \begin{bmatrix} -\Gamma_{11}^1\Gamma_{11}^{-1}(\Gamma_{11}^1)^T & 0 & \Gamma_{11}^1\Gamma_{13} & \Gamma_{11}^1\Gamma_{14} \\ * & -\Gamma_{22} & \Gamma_{23}^1\Gamma_{23} & \Gamma_{23}^1\Gamma_{24} \\ * & * & -M_{asi} & -\tilde{E}_{1asii}^T\mathcal{S} \\ * & * & * & \Gamma_{44} \end{bmatrix} < 0. \tag{12.32}$$

Considering $\bar{P}_{at} > 0$, it follows that $(\bar{P}_{at} - N_s)\bar{P}_{at}^{-1}(\bar{P}_{at} - N_s)^T > 0$. Furthermore, we have

$$\bar{P}_{at} - N_s - N_s^T > -N_s \bar{P}_{at}^{-1} N_s^T. \tag{12.33}$$

Accordingly, it yields that

$$\Gamma_{11} - \Gamma_{11}^1 - (\Gamma_{11}^1)^T > -\Gamma_{11}^1 \Gamma_{11}^{-1} (\Gamma_{11}^1)^T. \tag{12.34}$$

From (12.32), it follows that

$$\Gamma''_{asiit} = \begin{bmatrix} \Gamma_{11} - \Gamma_{11}^1 - (\Gamma_{11}^1)^T & 0 & \Gamma_{11}^1 \Gamma_{13} & \Gamma_{11}^1 \Gamma_{14} \\ * & -\Gamma_{22} & \Gamma_{23}^1 \Gamma_{23} & \Gamma_{23}^1 \Gamma_{24} \\ * & * & -M_{asi} & -\breve{E}_{1asii}^T S \\ * & * & * & \Gamma_{44} \end{bmatrix} < 0. \tag{12.35}$$

On the other hand, define

$$\begin{cases} \hat{\mathscr{A}}_{sj} = X_s \hat{A}_{sj}, \ \hat{\mathscr{E}}_{sj} = \hat{E}_{sj}, \\ \hat{\mathscr{B}}_{sj} = X_s \hat{B}_{sj}, \ \hat{\mathscr{F}}_{sj} = \hat{F}_{sj}, \ i = j, \end{cases} \tag{12.36}$$

and

$$N_s = \begin{bmatrix} N_{1s} & X_s \\ N_{2s} & X_s \end{bmatrix}. \tag{12.37}$$

Then based on (12.29), we can clearly find that

$$\Lambda_{asiit} = \Gamma''_{asiit}. \tag{12.38}$$

Based on (12.32), (12.34) and (12.38), it is easy to find that we can obtain (12.15) from (12.29). Adopting the similar method to (12.16) and (12.30), we can also achieve (12.16) from (12.30). And the filter matrices can be derived from (12.36). The proof is completed.

On the basis of Theorem 12.2, another filter design method is developed by using Finsler's Lemma.

Theorem 12.3 *If there exist n-dimensional matrices $P_{ai} = \begin{bmatrix} P_{1ai} & P_{2ai} \\ * & P_{3ai} \end{bmatrix} > 0$, $M_{asi} = \begin{bmatrix} M_{1asi} & M_{2asi} \\ * & M_{3asi} \end{bmatrix}$ $(n = 2(n_{\tilde{x}} + n_x))$, $(n_{\tilde{x}} + n_x)$-dimensional matrices X_s, $\hat{\mathscr{A}}_{sj}$, $\hat{\mathscr{B}}_{sj}$, $\hat{\mathscr{E}}_{sj}$, and $\hat{\mathscr{F}}_{sj}$ for any $a \in \mathcal{V}$, $s \in \mathcal{L}$ and $i, j, t \in \mathcal{I}$ subject to*

$$\sum_{s=1}^{L} \varphi_{as} M_{asi} < P_{ai}, \tag{12.39}$$

$$Q_1^T \bar{\Lambda}_{asiit} Q_1 < 0, \tag{12.40}$$

$$O^T \bar{\Lambda}_{asiit} O < 0, \tag{12.41}$$

$$Q_2^T(\bar{\Lambda}_{asijt} + \bar{\Lambda}_{asjit})Q_2 < 0,\ i < j, \tag{12.42}$$

$$O^T(\bar{\Lambda}_{asijt} + \bar{\Lambda}_{asjit})O < 0,\ i < j, \tag{12.43}$$

where

$$\bar{\Lambda}_{asijt} = \begin{bmatrix} -\bar{\Lambda}_{11} & 0 & 0 & \bar{\Lambda}_{14} & \bar{\Lambda}_{15} \\ * & -\bar{\Lambda}_{11} & 0 & \Lambda_{24} & \Lambda_{25} \\ * & * & -\Lambda_{33} & \Lambda_{34} & \Lambda_{35} \\ * & * & * & -M_{asi} & \Lambda_{45} \\ * & * & * & * & \Lambda_{55} \end{bmatrix},$$

$$\bar{\Lambda}_{11} = \begin{bmatrix} \bar{P}_{1at} & \bar{P}_{2at} - X_s \\ * & \bar{P}_{3at} - X_s - X_s^T \end{bmatrix},$$

$$\bar{\Lambda}_{14} = \begin{bmatrix} \beta\hat{\mathscr{B}}_{sj}\bar{E}_{ai} & \hat{\mathscr{A}}_{sj} \\ \beta\hat{\mathscr{B}}_{sj}\bar{E}_{ai} & \hat{\mathscr{A}}_{sj} \end{bmatrix},\ \bar{\Lambda}_{15} = \begin{bmatrix} \beta\hat{\mathscr{B}}_{sj}\bar{F}_{ai} \\ \beta\hat{\mathscr{B}}_{sj}\bar{F}_{ai} \end{bmatrix},$$

$$Q_1 = \begin{bmatrix} Q_{11} & Q_{12}^1 \\ 0 & Q_{22} \end{bmatrix},\ Q_2 = \begin{bmatrix} Q_{11} & Q_{12}^2 \\ 0 & Q_{22} \end{bmatrix},\ O = \begin{bmatrix} 0 \\ O_{12} \end{bmatrix},$$

$$Q_{11} = \begin{bmatrix} 0 \\ I \\ 0 \end{bmatrix},\ Q_{12}^1 = \begin{bmatrix} 0\,0\,0 & A_{ai} & 0 & B_{ai} \\ 0\,0\,0 & 0 & 0 & 0 \\ 0\,0\,0 & 0 & 0 & 0 \end{bmatrix},$$

$$Q_{12}^2 = \begin{bmatrix} 0\,0\,0 & \frac{A_{ai}+A_{aj}}{2} & 0 & \frac{B_{ai}+B_{aj}}{2} \\ 0\,0\,0 & 0 & 0 & 0 \\ 0\,0\,0 & 0 & 0 & 0 \end{bmatrix},$$

$$Q_{22} = diag\{I, I, I, I, I, I\},\ O_{12} = diag\{I, I, I, I, I\},$$

there exists an $(n_{\tilde{x}} + n_x)$-dimensional dissipative FD filter with $(n_{\hat{x}} = n_{\tilde{x}} + n_x)$ in the form of (12.7), which can guarantee that the FD system (12.11) is stochastically stable with strict dissipativity. Moreover, we can obtain the filter gains via solving (12.31).

Proof Inequalities (12.29) and (12.30) can be rewritten as

$$\Lambda_{asiit} = \bar{\Lambda}_{asiit} + \bar{O}^T N_s' \bar{Q}_i + \bar{Q}_i^T N_s'^T \bar{O} < 0, \tag{12.44}$$

and

$$\begin{aligned} \Lambda_{asijt} + \Lambda_{asjit} = &\ \bar{\Lambda}_{asijt} + \bar{\Lambda}_{asjit} + \bar{O}^T N_s'(\bar{Q}_i + \bar{Q}_j) \\ &+ (\bar{Q}_i^T + \bar{Q}_j^T) N_s'^T \bar{O} < 0, \end{aligned} \tag{12.45}$$

where

$$\bar{Q}_i = \begin{bmatrix} -I & 0 & 0 & 0 & 0 & 0 & \bar{A}_{ai} & 0 & \bar{B}_{ai} \\ 0 & 0 & -I & 0 & 0 & 0 & 0 & 0 & 0 \end{bmatrix},$$

$$\bar{O} = \begin{bmatrix} I & 0 & 0 & 0 & 0 & 0 & 0 & 0 & 0 \\ 0 & I & 0 & 0 & 0 & 0 & 0 & 0 & 0 \\ 0 & 0 & I & 0 & 0 & 0 & 0 & 0 & 0 \\ 0 & 0 & 0 & I & 0 & 0 & 0 & 0 & 0 \end{bmatrix}, \quad N_s' = \begin{bmatrix} N_{1s} & 0 \\ N_{2s} & 0 \\ 0 & N_{1s} \\ 0 & N_{2s} \end{bmatrix}.$$

At the same time, Q_1, Q_2 and O are orthogonal complements of $\bar{Q}_i^T$, $\bar{Q}_i^T + \bar{Q}_j^T$ and $\bar{O}^T$, respectively. Hence, via Finsler's Lemma, we find that the feasibility of (12.44) (or (12.29)) is equivalent to the feasibility of (12.40)–(12.41). It is easy to conclude that the relationship between (12.45) (or (12.30)) and (12.42)–(12.43) is equivalent by similar proof. The proof is completed.

Remark 12.4 From Theorem 12.1, it can be easily observed that there are some product terms between Lyapunov matrix $\bar{P}_{at}$ and system matrices like $\breve{A}_{1asij}$ and $\breve{B}_{1asij}$. Via introducing the slack matrix N_s, these couplings are eliminated, as shown in Theorem 12.2. Nonlinear matrix inequalities in Theorem 12.1 become linear. Then based on Theorem 12.2, we utilize Finsler's Lemma to remove unnecessary matrices N_{1s} and N_{2s}, and achieve Theorem 12.3, which further reduce the number of unknown variables. However, the number of LMIs increases by $0.5VLr(r+1)$, which is a trade-off between less conservatism and computational burden.

12.4 Conclusion

In this chapter, we have studied the dissipative asynchronous FD problem for nonlinear MJSs with data dropouts via the T–S fuzzy technique. A sufficient condition has been developed to ensure the stochastic stability and the strictly dissipative performance of FD systems by applying the Lyapunov function approach. We have established two LMI-based methods for the existence of the dissipative asynchronous FD filter, which can be cast into a convex optimization problem.

References

1. Wei, Y., Qiu, J., Karimi, H.R.: Reliable output feedback control of discrete-time fuzzy affine systems with actuator faults. IEEE Trans. Circuits Syst. I: Regul. Pap. **64**(1), 170–181 (2017)
2. Kommuri, S.K., Defoort, M., Karimi, H.R., Veluvolu, K.C.: A robust observer-based sensor fault-tolerant control for PMSM in electric vehicles. IEEE Trans. Ind. Electron. **63**(12), 7671–7681 (2016)
3. Rathinasamy, S., Karimi, H.R., Joby, M., Santra, S.: Resilient sampled-data control for Markovian jump systems with adaptive fault-tolerant mechanism. IEEE Trans. Circuits Syst. II: Express Briefs **64**(11), 1312–1316 (2017)

4. Li, H., Gao, Y., Wu, L., Lam, H.K.: Fault detection for T-S fuzzy time-delay systems: delta operator and input-output methods. IEEE Trans. Cybern. **45**(2), 229–241 (2015)
5. Su, X., Shi, P., Wu, L., Song, Y.-D.: Fault detection filtering for nonlinear switched stochastic systems. IEEE Trans. Autom. Control **61**(5), 1310–1315 (2016)
6. Chadli, M., Abdo, A., Ding, S.X.: H_{-}/H_{∞} fault detection filter design for discrete-time Takagi-Sugeno fuzzy system. Automatica **49**(7), 1996–2005 (2013)
7. Li, H., Chen, Z., Wu, L., Lam, H.-K., Du, H.: Event-triggered fault detection of nonlinear networked systems. IEEE Trans. Cybern. **47**(4), 1041–1052 (2017)

Index

S. Dong et al., *Control and Filtering of Fuzzy Systems with Switched Parameters*, Studies in Systems, Decision and Control 268,
https://doi.org/10.1007/978-3-030-35566-1

CPSIA information can be obtained
at www.ICGtesting.com
Printed in the USA
LVHW082105141220
674148LV00001B/31

9 783030 355685